Agriculture and Water Management

Agriculture and Water Management

Dr. P. Verma

AMIGA PRESS. INC
Delhi -110051

AMIGA PRESS. INC
H.O :- 28-A, Indra Park, Chander Nagar
Delhi :- 110051 (India)
Tel :- 011-22504028, +91-98736-57000
Email :- amigapress@gmail.com

This edition is published by Amiga Press.Inc

ISBN : 978-81-929542-1-9

First Published 2018

PRINTED : AMIGA PRESS INDIA

Preface

India has made lot of progress in agriculture since independence in terms of growth in output, yields and area under many crops. It has gone through a green revolution, a white revolution, a yellow revolution and a blue revolution. Today, India is the largest producer of milk, fruits, cashewnuts, coconuts and tea in the world, the second largest producer of wheat, vegetables, sugar and fish and the third largest producer of tobacco and rice. The per capita availability of foodgrains has risen in the country from 350 gm in 1951 to about 500 gm per day now, of milk from less than 125 gm to 210 gm per day and of eggs from 5 to 30 per annum despite the increase in population from 35 crores to 95 crores. However, India would have been in an even better position now both in terms of agricultural output and economic development had our planners given the required importance to its development it deserved in the early years since independence.

The agricultural produce in the country is freely marketed through private trade operating in organized markets/mandies. The role of the government normally is limited to protecting the interests of producers and consumers, only in respect of wage goods, mass consumption goods and essential goods. The government is promoting organized marketing of agricultural commodities in the country.

Agriculture in India is one of the most prominent sectors in its economy. Agriculture and allied sectors like forestry, logging and fishing accounted for 18.6% of the GDP in 2005 and employed 60% of the country's population. It accounts for 8.56% of India's exports. About 43% of India's geographical area is used for agricultural activity.

Despite a steady decline of its share in the GDP, agriculture is still the largest economic sector and plays a significant role in the overall socio-economic development of India. The monsoons play a critical role in the Indian sub-continent's agriculture in determining whether the harvest will be bountiful, average, or poor in any given year. The entire rainfall in the sub-continent is concentrated in the few monsoon months.

Integrated water resources management is the practice of making decisions and taking actions while considering multiple viewpoints of how water should be managed. These decisions and actions relate to situations such as river basin planning, organization of task forces, planning of new capital facilities, controlling reservoir releases, regulating floodplains and developing new laws and regulations.

The need for multiple viewpoints is caused by competition for water and by complex institutional constraints. The decision-making process is often lengthy and involves many participants.

Shortage of water available for agriculture is a problem in most parts of the world. Shortage and degradation of agricultural land also greatly limit agricultural development in many regions. There is additionally increasing concern that agricultural development contributes significantly to general environmental degradation. These problems have reached dangerous levels in many countries such that economic development and social structures are threatened. If future food production is to keep pace with developing population, and natural environments are to be protected, then agricultural practice must change. An increased in the efficiency with which water is used in agriculture is a major priority. For many countries, agriculture is the largest water user and to improve water use efficiency in agriculture would have a significant impact on sustainable development.

This book covers all the aspects of agriculture. It is hoped, the book will be found to be of immense value to the students of higher education, researchers and teachers.

—*Editor*

Contents

1

Agriculture in India

India has made lot of progress in agriculture since independence in terms of growth in output, yields and area under many crops. It has gone through a green revolution, a white revolution, a yellow revolution and a blue revolution. Today, India is the largest producer of milk, fruits, cashewnuts, coconuts and tea in the world, the second largest producer of wheat, vegetables, sugar and fish and the third largest producer of tobacco and rice. The per capita availability of foodgrains has risen in the country from 350 gm in 1951 to about 500 gm per day now, of milk from less than 125 gm to 210 gm per day and of eggs from 5 to 30 per annum despite the increase in population from 35 crores to 95 crores. However, India would have been in an even better position now both in terms of agricultural output and economic development had our planners given the required importance to its development it deserved in the early years since independence.

Even today, the farmers in India are able to obtain only 15 per cent of their requirements of agricultural credit from banks. The various state seed corporations are able to produce only 1 0 per cent of the seeds required by our farmers. At present only 2330 per cent of the farmers are able to derive any benefits of extension services provided by various government agencies and every year about 20 per cent of the crop is lost due to mishandling, spillage, floods, droughts and pests and diseases.

In fruits and vegetables the loss is around 30 per cent. The farmers are still being exploited by moneylenders who provide finance at exorbitant interest rates and there are cartels of traders who pay very little for their produce even in the well recognized mandies in the country.

TRENDS IN THE AGRICULTURE SECTOR

There were three distinct periods in the evolution of Indian agriculture and agricultural policy after Independence. The first period was before the Green Revolution up to the mid-1960s (Phase 1). The second period was the

Green Revolution from the mid-1960s till the late 1980s (Phase 2). The third phase continues from then to the present day (Phase 3).

The first period saw the government focusing policy on the following:

- irrigation,
- land reforms,
- community development, and
- restructuring of rural credit institutions.

Land reforms involved abolition of intermediaries, tenancy reforms, acquisition of surplus land through ceilings on holdings and redistribution and consolidation of holdings; success varied across states. Land reforms quite successfully altered agrarian structures in states like Kerala, West Bengal and Karnataka, but failed to do so in states such as Bihar, Orissa and Rajasthan. Changes in agrarian structures occurred in Punjab, Haryana, Western Uttar Pradesh, Tamil Nadu and Andhra Pradesh.

A Community Development Programme was started in 1952, for integrated village development by coordinating the development of agriculture, animal husbandry, infrastructure and extension at the block level. Also in 1952 the National Extension Programme was spliced into the Community Development Programme, incorporating provision of technical inputs. The Intensive Agricultural District Programme (IADP), started in selected potential districts in 1960 and 1961, aimed to provide a package of high-yielding inputs (seed, fertilizers, plant protection measures, etc.) to farmers. The Intensive Agricultural Area Programme (IAAP), started in selected potential blocks in 1964 and 1965 aimed to provide technological inputs for identified crops.

Whether these changes were successful or not depends upon how one benchmarks success. Compared to the low or negative growth in the period immediately preceding Independence, they were successful. Both productivity and output increased during most of the period. However, progress in productivity improvements was not commensurate with the requirements even during good monsoon years. There were institutional constraints in providing inputs like high-yielding varieties, fertilizers and irrigation, coupled with a switch during the Second Plan (1956–1961), with reduced investments in agriculture as fallout. In other words, the government was unable to inject the required financial resources, or develop adequate institutional back-up to enable widespread and sustained implementation. Moreover, the government was also unable to effect improvements in technology and the usage of improved seeds, inputs and methods did not occur until much later.

On the positive side the first phase did succeed in other aspects that facilitated later changes. Land reforms (although limited to some states and discussed elsewhere), assigning ownership rights to tenants, consolidation of holdings, among others, were important measures that started during this period and have continued to a varying extent to the present day. The establishment of many agricultural technology institutions and the birth of a widespread extension services system also transpired during this time (although it is now ineffective in many states). Most importantly, administrative systems were developed during this period with the capacity to take on the challenge of the Green Revolution when the technologies finally became available. The mid-1960s witnessed a succession of exogenous negative shocks that had a major impact on agricultural output: a war with China in 1962, a war with Pakistan in 1965 and successive monsoon failures in 1965 to 1966 and 1966 to 1967. India had to resort to large-scale imports of food.

The Green Revolution package was based on increased availability/ access to:

- high-yielding varieties (HYV) of seeds,
- fertilizers,
- irrigation,
- biochemicals,
- extension services,
- availability of credit,
- establishment of marketing and price-support mechanisms for farmers.

This was spliced with a price-support policy. The government's role in ensuring the availability of complementary inputs for HYV seeds was quite significant. Ahluwalia summarizes it well, " it required a comprehensive strategy for agricultural change requiring active Government intervention in many dimensions...".

A sustained effort was needed to expand irrigation with a shift from major to medium and minor irrigation projects. It was necessary to move the banking system into rural areas to provide credit for the purchase of biochemical inputs needed for HYVs. Nationalized banks were given the task of upgrading their rural operations and they succeeded to a large extent. Primary agriculture markets were regulated and some of the usurious practices of traders stopped. Extension services were set up and backed by numerous agricultural research establishments. Development of appropriate varieties was critical given the heterogeneity in agroclimatic conditions. The ability of the government to effect adequate coordinating mechanisms was also quite important. Provision of credit was facilitated through the nationalization of banks. Fertilizer availability and accessibility was expanded through subsidies and was juxtaposed by public and private sector expansion in manufacturing. Moreover price support was instituted at remunerative prices. This coordination role cannot be understated. While subsidized inputs were becoming available and productivity increases were rapid and concentrated, the price-support mechanism ensured that incomes for farmers increased. The three key characteristics of the mandate were: simultaneous handling of constraints in supply and marketing, a coordinated approach and recognition of heterogeneity in agroclimatic conditions.

This approach can be assigned most of the credit for the success of the Green Revolution. The growth rates in area, production and productivity of major crop groups. The asterisks in the table indicate the major contributing factors behind increase in production during that period. From 1949–1950 to 1964–1965, non-foodgrain was the major component of crop production and for both foodgrain and non-foodgrain, production growth primarily transpired through area expansion. This was reversed from 1967–1968 to 1980–1981. During this period, expansion of production for both foodgrain and non-foodgrain occurred through productivity increases, area expansion having slowed down. This trend continued from 1979–1980 to 1989–1990, with productivity increases becoming sharper and the area under foodgrain actually declining. However, in the last few years, productivity growth has slackened, and so has growth in production, particularly of foodgrain.

However the Green Revolution had several positive aspects: (1) As foodgrain production increased faster than the population growth rate, per capita availability of food increased. Per capita net availability of cereals and pulses was 394.9 grams in 1951, 468.7 grams in 1961, 468.8 grams in 1971, 454.8 grams in 1981, 510.1 grams in 1991 and 416.2 grams in 2001 (GOI 2004–2005). (2) Per capita income generation in agriculture increased. (3) Agriculture was better protected from the ravages of drought. (4) There was greater commercialization and diversification, with cropping patterns changing in favour of commercial crops and moving away from coarse cereals, even for small and marginal farmers. (5) Capital accumulation rose, including investments from the private sector. (6) While there may have been some reservations about the initial distributional impact of the Green Revolution package, the benefits became more broad-based in subsequent years.

Many holdings are marginal (less than 1 ha) or small (between 1 and 2 ha). Barring pulses and coarse cereals, all other crops seem to have benefited from the new technologies. In addition, there was a paradigm shift from foodgrain towards commercial crops and even fruits and vegetables, this diversification being at the expense of pulses and coarse cereals. Although a large percentage of small and marginal farmers is still dependent on foodgrain, there was some diversification among small and marginal farmers. For instance, the share of cereals in area cultivated by small and marginal farmers declined from 71.44 percent in 1970 to 1971 to 70.57 percent in 1980 to 1981 and 66.22 percent in 1990 to 1991; the share of fruit and vegetables increased in area from 2.43 percent in 1970 to 1971 to 3.25 percent in 1980 to 1981 and 3.71 percent in 1990 to 1991 (Despande *et al.* 2004). Thus it can be argued that there has been a degree of commercialization and diversification among small and marginal farmers but this could have been greater.

However there are some qualifiers. The first is that the coordinated approach did not lead to the spread of the benefits of the new technologies across India. Only a few areas benefited. It is difficult for a central government to introduce measures and mechanisms that are fine-tuned to the requirements of each state and district. , Success was based on the term "provision" — *inter alia* provision of credit, HYVs, subsidized fertilizers and extension services. These had a negative long-term impact on the development of market-backed services and commodities. Lastly, the new technologies were biased towards high usage of water for irrigation. As many parts of the country were not irrigated, power-operated motorized tubewells sprung up across the country. Crops such as rice in Punjab and sugar cane in Maharashtra have become quite important even though surface water is limited in some of these areas. The result has been a rapid fall in water levels.

In the 1980s, when the balance between demand and supply of cereals (foodgrain) was in sight, the overall goal of agricultural strategy/policy shifted from "maximization of production of foodgrain" to "evolution of a production pattern in line with the demand pattern". The implication of this shift is that the emphasis of the policy shifted from foodgrain to other agricultural commodities like oilseed, fruit and vegetables. The shift helped to increase the output of non-cereal food items.

Diversification has been occurring in India since Phase 1, however, it became more prominent in the 1980s (but waned in the 1990s). There is little agreement on how much of this was due to the government efforts and how much transpired because of changing economic conditions. It would be fair to argue for oilseed that the shift from coarse cereals to oilseed in rain-fed areas was aided by (i) a protective trade environment; (ii) favourable price policy; and (iii) the connecting of the Technology Mission on Oilseeds. At the same time diversification was not merely attributable to greater area under high-value commodities; the improvements in yield also would have been generated by government-aided supply and technology-related factors.

The improvements in both production and yield for the crop sector. There were significant improvements for livestock in the 1980s — the value of the livestock sector in total agricultural output value increased from about 18 percent in the triennium ending (TE) 1980 to 1981 to 23 percent in TE 1990 to 1991 (although it remained stagnant at that level by TE 1997 to 1998. Almost 70 percent of the livestock sector's production (in value terms) is accounted for by the milk and products subsector; the role of "Operation Flood" cannot be understated.

The value of the fisheries subsector swelled by about 50 percent during the 1990s, although as a share of the total it declined from 1.3 to 1.0 percent of the total agriculture sector value during the period. There were significant and coordinated government efforts to increase fish production. The central government's outlays towards the fishery sector as a share of the total agriculture sector outlays were more than doubled from 2 to 3 percent in the 1970s to about 5.5 percent in the 1980s and 1990s. Production and development programmes were instituted in both marine and inland areas. Farmers' development agencies were established both for fresh and brackish water areas. These programmes included technology upgrading components, encouragement and involvement of the private sector in activities such as seed, feed and other inputs and also creation of suitable infrastructure for storage, transport, marketing and credit. By 1998/1999 more than 50 seed hatcheries at the national level had been established, fishery industrial estates had been created and 30 minor fishing harbours and 130 fish-landing centres had evolved.

Diversification in favour of high-value commodities is considered by some to have only partly benefited through government efforts. Both econometric and GIS-based studies also suggest that the presence of infrastructure aided diversification. Rao *et al.* (2004) found that in their sample, the areas close to large urban centres (and as a result with better connectivity to urban centres) could diversify more to high-value commodities. They concluded that urbanization is a strong demand driver for high-value commodities. In an earlier econometric study, Joshi *et al.* came to a similar conclusion — that diversification into high-value commodities was demand driven, unlike the supply driven Green Revolution.

Between 1990–1991 and 1998–1999, the annual average increase in yields was only 1.79 percent and 1.31 percent for wheat and rice, respectively. Despite the successes of the Green Revolution, there continue to be concerns with agricultural performance during Phase 3, the present phase of globalization and diversification, initiated in the 1990s, and some macro issues emerge.

The first macro point is the one already alluded to indirectly. Compared to the 1980s, there was deceleration of agricultural growth in the 1990s. The cuts across all crops and both production and yields. The quality of the system for collecting agricultural statistics and the extent of deceleration can be debated, but the conclusion that growth in the crop sector decelerated in the 1990s is fairly obvious. In any case, deficiencies in the quality of statistics are constant factors and cannot explain the deceleration. Indeed, growth in non-food and non-crop output was also faster in the 1980s than in the 1990s and growth in animal husbandry, poultry, dairy, horticulture and fisheries also slowed down. This deceleration happened despite the reversal of historical discrimination against agriculture. In the 1990s, the terms of trade moved in favour of agriculture and this was independent of the terms of trade measures used. There were several reasons for this reversal of trend. There were higher prices for rice and wheat through support/procurement policies and prices of other crops also increased. There is a positive correlation between procurement prices, open market prices and higher prices in the PDS (public distribution system). Simultaneously, because protection granted to manufacturing declined, the relative prices of manufactured products also declined. *A priori*, one should therefore have expected a positive supply response and increased capital accumulation.

However farm profitability declined in the 1990s. Not all farmers had access to higher support prices for rice, wheat or sugar cane, there was a deceleration in yields and real input prices also increased, thanks to prices for fertilizers, power and diesel that were closer to market-determined prices. Higher cereal prices also contributed to increases in wage costs.

Consequent to per capita income growth, NSS data show changes in consumption patterns, with a decline in consumption of cereals in both rural and urban India, despite cereals still being important in the food basket. There has been a shift from consumption of coarse cereals to rice and wheat and also a shift towards consumption of fruit and vegetables and even fish and eggs. Some of these changes. Consumption patterns are changing, even for the bottom 30 percent of the population and the shares of non-cereal food (fruit and vegetables) and non-food items have been increasing in consumption baskets. Increased cereal prices and lower rural incomes may have depressed demand for both cereal and non-cereal food, but there is no denying that preferences are also changing.

The Reason for Deceleration

Why did deceleration occur? More importantly, why did deceleration occur when better technologies were more readily available? In answering this question, it is important to distinguish between generic problems that continue to constrain agriculture and issues that became constraints in the reform decade of the 1990s. The former constitutes part of the agricultural cum rural reform agenda, the latter is more specific. First, one confronts a diminishing returns argument, with total factor productivity declining, as opposed to an argument based on increases in input costs or a slower increase in output prices. This argument is usually applied to Punjab, Haryana and the western parts of Uttar Pradesh where agriculture has become overcapitalized. In these traditional Green Revolution areas, there are also questions about unsustainable practices like excessive use of water and imbalanced use of fertilizers. Land has become degraded. These are traditional arguments associated with the Green Revolution and concern reduction in soil fertility, excessive use of fertilizers and imbalance of nutrient content in the soil, problems related to biomass availability, genetic erosion, waterlogging and salinization, depletion of groundwater tables, imbalances in nutrient availability because of changes in cropping patterns and contamination of waterbodies and soil by pesticides and fertilizers. Then there is the issue related to lowering of farm profitability in the 1990s. While these are important topics, they are not very convincing in explaining the 1990s deceleration, unless one plugs in a regional dimension. In the 1980s, availability of power, irrigation and infrastructure helped the spread of the Green Revolution to the Eastern Region, particularly for paddy rice. Owing to power shortages, among other factors, this osmosis was less evident in the 1990s.

Another explanation for the 1990s deceleration is reduced public investments, especially in irrigation; this is directly reflected in the reduced share of capital formation in agriculture in the overall GDP. Moreover, in the 1990s, the quality of public sector agricultural research, technology development and extension services deteriorated.

While public investments in agriculture have declined, especially investments in research, private investments have not been able to compensate for the decline in public sector investments. Thus the burden of fiscal reform has been borne by agricultural investments, especially at the state level, including expenditure on R&D. More specifically, expenditure on research stagnated, while that on extension declined. The state withdrew from spending on agriculture, without establishing alternative institutional mechanisms. State governments lack resources, a problem that was aggravated in the 1990s, and adversely affected the development of agricultural infrastructure.

This can be juxtaposed with the failure of risk mitigation instruments to develop and general regulatory failure. There were no coordinated policies or crop adjustment programmes to enable the move from rice and wheat towards pulses, oilseed or other crops. Diversification led to risk and uncertainty and in the absence of institutions, this was difficult to handle.

Within public sector institutions, there was deterioration in management and monitoring norms. Hence, there were instances of adulterated fertilizers and seeds that were substandard.

Whether extension departments of state governments are the best agencies for quality checks and inspections of inputs like fertilizers, insecticides, pesticides, feed and seeds is debatable. However, the quality of these extension services deteriorated in the 1990s.

While only tangentially related to the slowdown of the 1990s, there is the question of the form public investments in agriculture should take. The Steering Group on Agriculture and Allied Sectors (SGAAS) does not question this, but the Approach Paper to the Tenth Five Year Plan (APTFYP) does, "The policy approach to agriculture, particularly in the 1990s, has been to secure increased production through subsidies in inputs such as power, water and fertilizer, and by increasing the minimum support price rather than through building new capital assets in irrigation, power and rural infrastructure. This strategy has run into serious difficulties." The statement further identifies that the deterioration in state finances and the financial non-viability of the State Electricity Boards have meant that crowding out has occurred in public agricultural investment/expenditures for:

- roads,
- irrigation,
- expenditure on technological upgrading,
- maintenance of canals and roads,
- expansion of power supply to rural areas not covered,
- improvement in quality of rural power.

Moreover the combination of dependence on subsurface water and free/ subsidized power is not only environmentally harmful but has also led to:

- excessive use of water that produced waterlogging in many areas,
- falling water tables.

At the same time, there is no evidence of any improvement in income distribution. Last, it is also increasingly being noticed that despite the best intensions of the policy there is a significant imbalance between usage of N, P and K fertilizers. Moreover, financial constraints are increasing both for the state governments and their State Electricity Boards (SEBs).

These problems are particularly severe in the poorer states. The equity, efficiency and sustainability of this approach are questionable. The subsidies have grown in size and are now financially unsustainable. Power continues to be supplied free for farmers in many states and is otherwise highly subsidized given the high cost of production of SEBs. According to one estimate energy subsidies (of which power is one component) accounted for about 10 percent of all the subsidies in the late 1990s. This figure is likely to increase subsequently.

Moreover, the small and marginal farmers have been less able to access credit despite the existence of a large rural credit system. Poor land records are endemic across rural India. Also much of the agriculture sector is in the unorganized segment.

But the financial system is largely in the organized sector and requires credible land ownership records, systematic record keeping and formal contracts. This has contributed to a situation where the bulk of the small and marginal farmers cannot access formal credit institutions. In the 1990s, availability of credit for large farmers may have risen, but declined for small farmers. Moreover, although 18 percent of priority sector lending is supposed to be for agriculture, agriculture's share in net bank credit has been more like 12 percent.

In summary, there is unarguable stagnation in Indian agriculture and more so in the foodgrain sector.

This is due to a combination of factors ranging from overdependence on the government in some areas/sectors, falling ability of the various government entities to subsidize and/or directly provide various inputs and services and overuse of resources. The "mission mode" that was so successful in tackling the problems of low production cannot be a vehicle to take on this set of problems.

This requires efficient functioning of the price mechanism, smooth transactions between the various stakeholders, the alignment of incentives at the microlevel and fine-tuning of technology to the needs of the farmer. There are arguments for and against the current system of minimum support price (MSP) and subsidized inputs.

AGRICULTURE - SUB-SECTORS

Indian agriculture broadly consists of four sub-sectors. Agriculture proper including all food-crops oilseeds, fiber, plantation crops, fruits and vegetables is the largest accounting for nearly 70 percent of the agriculture sector as a whole. The rapid growth in this sub-sector through exploitation of wastelands and fallows, spread of irrigation and adoption of production enhancing technologies was critical in transforming India from a country vulnerable to food shortages to one of exportable surplus. Although this sub-sector has made impressive progress its share in the sector as a whole has declined from 78 percent in 1960-61 to less than 70 percent by early 90s.

Table. Agriculture sub-sectors – value of output (%)

Sub-sector	1960-61	1970-71	1980-81	1993-94	2014-15
Agriculture	78.0	79.3	75.3	68.5	70.0
Livestock	16.6	15.1	17.2	24.5	25.0
Forestry	4.4	4.4	5.9	3.8	3.0
Fisheries	1.0	1.2	1.6	3.2	2.0
	100.0	100.0	100.0	100.0	100.0
Total value at Current price (Rs. Billion)	89	218	615	2740	2740

Correspondingly the share of livestock sector has increased considerably. The livestock industry has grown from Rs. 15 billion in early 1960s to Rs. 100 billion by 1980-81 and Rs. 672 billion by 1993-94. In nominal terms the sector grew at almost 15 percent per annum during 1980s.

Milk production, which was almost stagnant for two decades ending 1970, grew by over 5 percent per annum in the 80s. Similarly, production of eggs increased at the rate of about 6.5 percent during the same period. As a result the share of livestock increased from about 17 percent till early 80s to 25 percent by 1993-94.

Though it plays relatively a minor role within the sector as a whole, fishing sub-sector activities have been on the rise. The sub-sector has grown from only Rs. 3 billion in 1970-71 to nearly Rs. 90 billion in 1993-94.

The growth was particularly rapid in 70s and 80s. Value added increased at over 5 percent per annum during this period. In real terms forestry and logging activities have been on the decline since mid seventies. As of 1993-94, the size of the industry in terms of value of output was 103 billion.

Over the past three decades, the country has successfully transformed itself from a food deficit economy to one which is essentially self sufficient in availability of foodgrains and other essential commodities, albeit only at the prevailing level of effective demand.

Annual aggregate foodgrains production, which averaged about 82 million tonnes in 1960-61 increased to 123.7 and 172.5 million tonnes for the trienniums ending 1980-81 and 1990-91 respectively.

Current (1998-99) production level is 195 million tonnes and the country has been able to accumulate substantial, (35 million tonnes) stocks of foodgrains to cope up with any sudden difficulties arising from drought or a similar situation in any part of the country.

Table. Foodgrains Production

Year	Area (million ha)	Production (million tonnes)	Yield kg/ha
1950-51	97.32	50.82	522
1960-61	115.58	82.02	709
1970-71	124.32	108.42	872
1980-81	126.67	129.59	1023
1990-91	127.84	176.39	1379
1998-99	124.0	195.25	1574
2014-15	124.0	197.25	1600

Increased outputs, have been achieved chiefly by adopting, since mid sixties, a strategy aimed at increasing foodgrains production by concentrating public sector efforts and resources in regions with a high potential for quick and substantial productivity gains through increased cropping intensity and average yields. These were the areas favoured by agroclimatic resource conditions and where irrigation facilities already existed or could be developed relatively rapidly. The main elements of this strategy were:

(i) expansion of irrigation coverage,

(ii) increased provision and utilization of key inputs – mainly high yielding varieties (HYVs) of crops, mainly of wheat and rice and chemical fertilizers and plant protection chemicals,

(iii) expansion and improvement of institutional support services such as research and extension and

(iv) price policies favourable to producers of major foodgrains.

The success of this strategy was made possible by development and availability of replicable production technology packages, so called 'Green Revolution' technologies. Irrigation facilitated double cropping and widespread adoption of HYVs. The HYVs performed particularly well under irrigated conditions, were highly responsive to fertilizers and their short duration permitted increases in cropping intensities.

Irrigation development was the cornerstone of the strategy. Undivided India was amongst the largest irrigated areas in the world. With partition nearly one-third of the irrigated area went to Pakistan. At the time of independence the net irrigated area was 20.9 million ha (gross irrigated area 22.6 million ha). Recognizing large-scale development of irrigation facilities as critical to rapid agricultural growth, the country has spent about Rs. 45,000 crores on irrigation development in the first four decades after independence. During the period 1950-51 to 1965-66 development of irrigation through government canals grew from 7.2 million ha to 9.8 million ha – a growth rate of 2.1 percent per annum. During 1970s this pace dropped slightly to

1.9 percent. In 1980s the rate of increase dropped significantly to 1.1 percent per annum.

The growth of tube-well irrigation, however, increased rapidly from 4.5 million ha in 1970-71 to 9.5 million ha in 1980-81 and then to 14.3 million ha by 1990-91. The net irrigated area increased from 31 million ha in 1970-71 to 53.5 million ha in 1995-96 which corresponds to 22 percent of the net sown area in 1970-71 and 37.63 percent in 1995-96 *(Tables 3, 4)*. With improvements in irrigation efficiency the gross irrigated areas has increased to 71.51 million ha. The percentage of gross cropped area service by irrigation increased from 18.3 percent in 1960-61 to 23.0 percent in 1970-71 and to over 38 percent at present.

Table. India – irrigated area

Year	Net	Gross	Net as % of net sown	Gross as % of gross cropped
	––– million ha	–––	–––– %	––––
1970-71	31.10	38.19	22.17	23.04
1980-81	38.72	49.78	27.66	28.83
1990-91	48.02	63.20	33.61	34.03
1995-96	53.51	71.51	37.63	38.33
2010-11	56.00	72.00	38.00	39.00

Table. Sources of irrigation

Year	Net Irrigated Area (million ha)	Sources				
		Canals	Tanks	Tube-wells	Other wells	Other Sources
		(% net irrigated area)				
1960-61	24.80	10.4	4.6	0.2	7.2	2.4
1970-71	31.10	41.28	13.22	14.34	23.88	7.29
1980-81	38.72	39.49	8.22	24.62	21.08	6.59
1990-91	48.02	36.34	6.13	29.69	21.73	6.11
1995-96	53.01	32.04	5.81	33.52	22.16	6.47
2010-11	55.00	33.00	6.00	34.00	23.00	6.66

Fertilizers have constituted yet another key input in addition to expanded irrigation and spread of HYVs in achieving goals of high production and productivity. India currently occupies third position in the world, after China and USA, in terms of fertilizer production and consumption. Consumption of fertilizers has increased from 1.54 million tonnes in 1967-68, representing the pre green revolution era, to 17.31 million tonnes currently (1997-98).

The average per hectare use of fertilizers currently around 85 kg per hectares is the lowest among several Asian countries. However, rice and wheat account for a major fraction, around 65 percent of the total fertilizer consumed in the country, with very little fertilizers going to the rainfed areas. According to some current projections, fertilizer's use will need to increase to 30-35 million tonnes to meet the foodgrains need of 2020. The demand for nutrients will stretch by almost another 15 million tonnes if requirements for horticulture, vegetables and plantation and commercial crops are included. At present domestic production of N and P fertilizers (13.42 million tonnes) falls short of consumption by over 20 percent. In addition the entire requirement of K fertilizer is imported. Most agricultural development programmes initiated in 1960s were concentrated in regions of high potential. Thus five states, Punjab, Haryana, Uttar Pradesh, Andhra Pradesh and Tamil Nadu account for 50 percent of the country's net irrigated and 53 percent of the gross irrigated area. The combination of expanding irrigation coverage and widespread adoption of short duration HYVs led to significant increases in cropping intensities. Acreage cropped more than once per year increased from 13 million ha in 1950-51 to about 44 million ha at present. Average cropping intensity for the country as a whole rose from 115 percent in 1960-61 to 131 in 1993-94. By 1993-94 cropping intensity has risen to 187 percent in Punjab, 167 percent in Haryana and 142 percent in Uttar Pradesh.

Table. Cropping intensity

Years	Net sown area	Gross cropped area	Area cropped more than once	Cropping intensity (%)
	million ha			
1950-51	118.8	131.9	13.1	111.1
1960-61	133.2	152.8	19.6	114.3
1970-71	140.3	165.8	25.5	118.2
1980-81	140.0	172.6	32.6	123.3
1990-91	143.0	185.7	42.7	129.9
1993-94	142.0	186.4	44.4	131.2
2010-11	165.0	190.0	45.0	132.6

An important consequence of the strategies adopted since sixties has been to boost production of, chiefly, two crops rice and wheat. Their share in total foodgrains production went up from 57 percent in 1970-71 to more than 75 percent in 1990-91. Production of foodgrains other than rice and wheat did not increase significantly and in the eastern region even the yield of rice did not increase. Agricultural production and income rose substantially in the north-western states of Punjab, Haryana, Western Uttar Pradesh, parts of Rajesthan, Tamil Nadu and Andhra Pradesh. By contrast productivity and output growth have been modest in eastern and central India and in deccan plateau. Progress was particularly slow in rainfed areas, which account for over 60 percent of the cropped area and where a great majority of rural poor are concentrated. An important impact of the strategies pursued in the 'Green Revolution' period has been intensification of regional disparities and imbalances in agricultural development and food availability and hence levels of food security. For the country as a whole while per capita availability of cereals has increased substantially, that of pulses has decreased significantly.

Table. Rice and Wheat Production Production (million tonnes)

Year	Rice	Wheat	Foodgrains	R+W as % of Foodgrains
	million tonnes			
1970-71	40.80	20.86	108.4	57.4
1980-81	49.91	34.55	129.6	65.4
1990-91	72.78	53.03	176.4	75.0
1996-97	81.74	69.35	199.4	75.0
2010-11	82.00	72.00	205.4	76.0

In summary an annual increase in foodgrains production of 3.22 percent during fifties was mainly because of expansion in area. Sixties recorded a low annual growth rate of 1.72 percent necessitating large-scale imports of foodgrains. Annual growth of 2.08 percent was recorded during seventies. This decade was the turning point in India's foodgrains economy leading to self sufficiency through significant productivity increase first in wheat and later in rice in the eighties. An annual growth of 3.5 percent in foodgrains in eighties was the hallmark of green revolution that enabled India to become self-sufficient and even a marginal exporter. The pace of growth slowed in nineties barely making or even slower than the population growth rate. This is a matter of concern.

Table. Annual compound growth rate of foodgrains production

Crop	1950-51 to 1959-60	1960-61 to 1969-70	1970-71 to 1979-80	1980-81 to 1989-90	1990-91 to 1997-98	2010-11 to 2013-14
Rice	3.28	- 8.05	1.91	4.29	1.53	2.00
Wheat	4.51	5.90	4.69	4.24	3.67	3.90
Coarse cereals	2.75	1.48	0.74	0.74	- 0.49	-0.50
Total cereals	3.00	2.51	2.37	3.63	1.84	1.90
Pulses	2.72	1.35	- 0.54	2.78	0.76	0.90
Total foodgrains	3.22	1.72	2.08	3.54	1.66	1.68

Table. Per capita net availability of cereals and pulses per day (g)

Year	Cereals	Pulses	Total
1951	334	61	395
1961	400	69	469
1971	418	51	469
1981	417	38	455
1991	468	42	510
1998	451	33	484
2010	485	34	489

CHANGING LAND-USE AND FUTURE OF AGRICULTURE

One of the most important consequences of growing pressure on land is the declining trend in the average farm size and the pattern of holdings.

According to the latest Agricultural Census in 1970-71 there were 70 million holdings operating 162 million ha. By 1990-91 there were 105 million holdings operating 165 million ha. The average farm size decreased from 2.30 ha in 1970-71 to 1.57 ha in 1990-91. As of 1990-91 about 78 percent of holdings were small (1.0 to 2.0 ha) and marginal (<1.0 ha). A little more than 20 percent of the farmers were semi-medium (2.0 to 4.0 ha) and medium (4.0 to 10.0 ha).

Large farmers (>10.0 ha) constituted only 1.6 percent of the total holdings. Over the twenty-year period since 1970 the proportion of marginal farmer has increased from 50 to 59 percent and that of large farmer has declined from about 4 to 1.6 percent.

Table. India – Land Use Categories (million ha)

Classification	1960-61	1970-71	1980-81	1990-91	1993-94
Reporting area	298.5	305.0	305.0	305.0	305.0
Area under non-agricultural uses	14.8	16.5	19.7	21.1	22.0
Barren and un-cultivable	35.9	28.2	20.0	19.4	19.0
Cultivable waste	19.2	17.5	16.7	15.0	14.5
Old fallow	11.2	8.8	9.9	9.7	9.7
Forests	54.1	63.9	67.5	67.8	68.4
Permanent pastures	14.0	13.3	12.0	11.8	11.2
Land under misc. trees etc.	4.5	4.3	3.6	3.8	3.7
Net sown area	133.2	140.3	140.0	143.0	142.1
Current fallow	11.6	11.1	14.8	13.7	14.3

The proportion of total area operated by marginal farmers increased from nine percent in 1970-71 to nearly 15 percent in 1990-91 while the proportion of large farmers declined from about 31 percent to 17 percent in the same period. The size of average holding is very unevenly distributed among the states. States with relatively large average size of operational holding are a mixed lot – they include states with large tracts of barren lands e.g. Rajasthan, Maharashtra and Madhya Pradesh on the one hand and agriculturally advanced state like Punjab on the other. These trends in farm size changes will have a profound effect on the future agricultural development strategies.

Table. Livestock and poultry population (million)

Category	**1982**	**1987**	**1992**
Cattle	192.45	199.69	204.53
Buffaloes	69.78	75.96	83.49
Sheep	48.76	45.70	50.79
Goats	95.25	110.20	115.28
Total (Livestock)	419.59	445.28	470.15
Total Poultry	207.74	275.32	307.10

Although India's population growth rate has slowed from 2.1 percent in 1980s to 1.8 percent in the 1990s and is expected to slow further in the coming decades, yet the population is projected to reach 1.33 billion by 2020 from the current one billion. The urban share of total population is projected to increase from 26 percent to 35 percent of the total population. Although incidence of poverty is falling, it is estimated that in 93-94 (upto which data is available) 320 million people constituting 36 percent of the population were below the officially defined poverty line.

The nature of the poverty line has been shifting. About 30 years ago 48.4 percent of those living in rural areas were poor and 20 percent of those living in the urban areas were classed as poor. Recent studies show that the number of poor in urban areas have been increasing at relatively higher rate compared to the rural areas. At present those below the poverty line in rural sector constitute 37 percent of the population while in the urban sector the percentage is 32 percent. In the context of poverty alleviation, therefore, emphasis will be required to be placed both on production of food by the poor as well as on the availability of food for the urban poor.

It needs to be recognized that a large proportion of the rural poor are located in regions of low potential for food production e.g. arid and semi-arid areas, hilly regions, degraded land and forest areas. Widespread hunger and malnutrition are the direct manifestation of poverty and will call for increasing efforts to produce more food at affordable price. Increasing population and economic growth are changing patterns of land use making potentially unsustainable demands on the country's natural resources.

Since early fifties the net area sown was expanded rapidly at first but at a diminishing rate since 1970 to reach approximately 142 million ha at present.

During 1950s and 1960s areas under agriculture expanded substantially as the fallows were reduced and cultivable wastes were put under the plough. The net area sown increased from 119 million ha in 1950-51 to 133 million ha by 1960-61 and further to 140 million ha by 1970-71. Fallow lands declined from 28 million ha in 1950-51 to 20 million ha by 1970-71. Cultivable wastelands declined from 23 to 17.5 million ha.

Land use intensity i.e. fraction of net sown area to total geographical area increase from 36 percent in 1950-51 to 40.5 percent in 1960-61 and 43 percent by 1970-71 where it has since stabilized.

- Cropping intensity i.e. gross sown area as percent of net sown area increased from 111 percent in 1950-51 to 115 percent in 1960-61, 118 percent in 1970—71 and 130 percent by mid 1990s.
- While the contribution of increased area in the growth of agriculture has declined over time, that of productivity has increased. The yield of all crops grew at 1.5 percent per annum between early 1950s and mid 1960s. the pace accelerated to 1.7 percent in the 1970s and then to 3 percent per annum between early 1980s and mid 90s. Unlike the gains in area, which benefited non-foodgrains, the gains in productivity accrued mostly to foodgrains.
- India's forest resources have been dwindling. According to the 'State of Forest Report' (1997) the total forest cover of the country is estimated at 63.34 million ha i.e. 19.27 percent of the geographic area of the country. Of these the dense forest (crown density more than forty percent) and open forest (crown density 10 to 40 percent) occupying about 11 and 8 percent of the geographic area respectively and mangroves occupy 0.15 percent of the geographic area. The country has lost about 5482 sq. km. of forest cover since the 1995 assessment. By any estimate the area under forest is far below the national policy goals and many areas nominally under forest are being used for non-forest purposes. Similarly 'uncultivated lands' such as permanent pastures, miscellaneous tree crops, cultivable wastes and fallow is subject to increasing competition from uses other than feeding livestock.
- The growth of livestock population is an important source of competition for land. The increase in number of major classes of livestock.

- The area sown to fodder crops is not recorded. Information available from other sources provide an estimate ranging from 4 to 5.5 percent of the net sown area and suggest that the area under fodder crops will have to increase to 10 percent or more to support increasing livestock based activity.

The pressure on India's land and water resources is seriously threatening native plant and animal diversity. India has uniquely rich and diverse genetic base. With increasing agriculture and economic development the genetic pool is declining. This decline, if unchecked and poorly managed can have unforeseen and adverse consequences for the sustainability of agriculture of the region.

AGRICULTURE IN THE CHANGING GLOBAL SCENARIO

Steady globalization of trade has profound implications for future agricultural development. The diversity of India's agro-ecological setting, high bio-diversity and relatively low cost of labour provide potential for agricultural competitiveness in a globalized economy. It is expected that with increasing globalization of markets over the years there will be demands for agricultural intensification. This will also be favoured because of greater backward and forward linkages between agriculture and food industry.

Therefore, increase in production and productivity are bound to be strategically important to economy. Intensification will not only favour alleviation of rural poverty but will also improve resource conservation particularly in the small farming sector where farmers can be encouraged to take up organized production of high value crops such as fruits, specialty vegetables, flowers medicinal and aromatic herbs etc. Stronger demands for crops of the small farmers' will not only improve incomes and welfare but will also make investments in technology and resource conservation more attractive.

The General Agreement on Tariff and Trade (GATT) and liberalization of global trade is bound to have impact on future land use and production pattern. Understanding the local, national and international environment under which agricultural production is taking shape will be crucial in developing our own strategies.

EXTENSION STRATEGIES

Since early fifties a number of public by funded agricultural development programmes have been sponsored. These have included programmes like the National Extension Service (NES) Blocks in 1953, the Intensive Agricultural District Programme (IADP) in 1961-62, the Intensive

Agricultural Area Programme (IAAP) 1964-65, the High Yielding Variety (HYV) programme 1966-67 and the Small and Marginal Farmers' Development Programmes (SMFDP) in 1969-70. Though these programmes had a perceptible impact the efforts did not get replicated over different areas and categories of farmers. In mid seventies based on pilot level project in Rajasthan Canal and Chambal command area a 'Training and Visit' (T&V) system of extension was promoted in different states. Extension efforts of the Indian Council of Agricultural Research through its research Institutes and the State Agricultural University were largely limited to demonstration of new technologies through such programmes as National Demonstration Project, Operational Research Project, the Lab to Land Programme and the Krishi Vigyan Kendras. However, there appears much to be desired in the way that extension programmes are conceived and implemented.

At present extension programmes are implemented in largely a top-down fashion leaving little scope for localized planning and action. Farmers are almost passive receivers and their involvement in the process of technology generation and adoption is almost absent. Extension services, at present, are almost exclusively in the public sector domain and there is no effort or institutional support for other operators e.g. the NGOs, the corporate bodies etc.

Extension programmes sponsored by the government operate largely in isolation and there appears a strong need to view the extension programmes as an integral part of the research and development process. The challenges facing agricultural development call for fundamental changes in our approach to technology transfer/extension programmes. Changes are necessary in the context of changing economic environment following policy adjustments in relation to privatization, deregulation and globalization calling for greater efficiency and effectiveness of the extension system.

More importantly there is need for

- Greater emphasis on providing producers with knowledge and understanding needed to overcome the problems or to exploit opportunities of their own specific production systems. Correspondingly there will be a need to de-emphasize 'package of practices' or the blanket recommendations, top down approach followed thus far.
- Shift in the focus of public extension systems from promoting inputs use to one on sustainable management of resources and improvements in the production system as a whole.
- Closer interaction between farmers, extension scientists and production system researchers in diagnosing problems and

identifying location specific recommendations emphasizing participation and education rather than being prescriptive.

- Widening the range of extension delivering agencies. While the publicly operated extension systems will continue to be important, there will appear a greater role for NGOs, farmers' associations and corporate sectors in particular situations. Role of commercial suppliers of seeds, agrochemicals, machinery, vaccines and medicines in providing advisories, as is already being done in a limited way, will need to be encouraged and factored into public system's own priorities.
- Wider and more creative use of mass media in tune with current developments in information technology to get information across to the farming community whose ability to overcome constrains at farm level will increasingly depend on access to reliable and up-to-date information.

TECHNOLOGICAL NEEDS AND FUTURE AGRICULTURE

It is apparent that the tasks of meeting the consumption needs of the projected population are going to be more difficult given the higher productivity base than in 1960s. There is also a growing realization that previous strategies of generating and promoting technologies have contributed to serious and widespread problems of environmental and natural resource degradation. This implies that in future the technologies that are developed and promoted must result not only in increased productivity level but also ensure that the quality of natural resource base is preserved and enhanced. In short, they lead to sustainable improvements in agricultural production.

Productivity gains during the 'Green Revolution' era were largely confined to relatively well endowed areas. Given the wide range of agroecological setting and producers, Indian agriculture is faced with a great diversity of needs, opportunities and prospects. Future growth needs to be more rapid, more widely distributed and better targeted. Responding to these challenges will call for more efficient and sustainable use of increasingly scarce land water and germplasm resources.

Technical solutions required to solve problems will be increasingly location-specific and matched to the huge agroecological/climatic diversity. Detailed indigenous knowledge and greater skills in blending modern and traditional technologies to enhance productive efficiency will be more than ever before, key to the farming success and sectoral growth. Most technological solutions will have to be generated and adapted locally to make them compatible with socio-economic conditions of farming community.

New technologies are needed to push the yield frontiers further, utilize inputs more efficiently and diversify to more sustainable and higher value cropping patterns. These are all knowledge intensive technologies that require both a strong research and extension system and skilled farmers but also a reinvigorated interface where the emphasis is on mutual exchange of information bringing advantages to all. At the same time potential of less favoured areas must be better exploited to meet the targets of growth and poverty alleviation. These challenges have profound implications for products of agricultural research. The way they are transferred to the farmers and indeed the way research is organized and conducted. One thing is, however, clear – the new generation of technologies will have to be much more site specific, based on high quality science and a heightened opportunity for end user participation in the identification of targets. These must be not only aimed at increasing farmers' technical knowledge and understanding of science based agriculture but also taking advantage of opportunities for full integration with indigenous knowledge. It will also need to take on the challenges of incorporating the socio-economic context and role of markets. With the passage of time and accelerated by macro-economic reforms undertaken in recent years, the Institutional arrangements as well as the mode of functions of bodies responsible for providing technical underpinning to agricultural growth are proving increasingly inadequate. Changes are needed urgently to respond to new demands for agricultural technologies from several directions. Increasing pressure to maintain and enhance the integrity of degrading natural resources, changes in demands and opportunities arising from economic liberalization, unprecedented opportunities arising from advances in biotechnology, information revolution and most importantly the need and urgency to reach the poor and disadvantaged who have been by passed by the green revolution technologies.

Another important implication of increasing globalization relates to the need for greater attention to the quality of produce and products both for the domestic and the foreign markets. This would imply that production must be tuned to actual rapidly changing product demand. Such adaptation to global markets would require state of the art research, which can be achieved only by setting global standards of research, focus on well defined priorities and mechanisms which permit close interaction of farmers with researchers, the private sector and markets.

DEVELOPMENT OF INDIAN AGRICULTURAL SECTORS

An appreciable pace of research has taken place in all sectors of agriculture including crops, horticulture, natural resource management, livestock, fisheries and agricultural engineering. The technology-led developments in agriculture have made India self-sufficient in foodgrains and a leading producer of several commodities in the world. The green revolution in crops, yellow revolution in oilseeds, white revolution in milk production, blue revolution in fish production and a golden revolution in horticulture bear an ample testimony to the contributions of agricultural research and development efforts undertaken in the country.

India has received worldwide acclaim in the field of agricultural research and education. The National Agricultural Research System (NARS), with the Indian Council of Agricultural Research (ICAR) as an apex body, has been striving for a holistic development of agriculture at the national level by planning, promoting, conducting and co-ordinating research, extension and education on all aspects of agriculture including animal sciences and fisheries for ensuring the optimal utilization of land, water, plant and animal genetic resources for sustainable agricultural development.

The Council undertakes research on fundamental and applied aspects in traditional and frontier areas to offer solutions to problems relating to conservation and management of resources, productivity improvement and health management of crops, animals and fisheries, post-harvest technology and value addition. This helps in developing new technologies in agriculture and allied sectors. The Council with its headquarters at New Delhi has a vast network of institutes all over the country consisting of 49 institutes including 5 National Institutes with 4 having Deemed University status, 5 National Bureaux, 33 National Research Centres, 11 Project Directorates and 78 All-India Co-ordinated Research Projects. Besides, it coordinates agricultural education in the country through a network of 34 State Agricultural Universities, 1 Central Agricultural University, 3 Academic Universities and 5 Deemed Universities. All activities under the transfer of technology programme of the Council have been integrated with Krishi Vigyan Kendras (KVKs) and the mandate enlarged to perform the function of on-farm testing and research, long-term vocational training, in-service training of grass-root level functionaries and frontline demonstrations. A network of 344 KVKs has so far been established in our country. A strong back-up support is provided through 8 Trainers' Training Centres to these KVKs by updating knowledge of the KVK trainers in the advancements made in research.

To strengthen research extension efforts, implementation of Institution-Village Linkage Programme (IVLP) at 70 centres in the country, establishment of 40 Agricultural Technology Information Centres (ATIC), strengthening of 53 Zonal Agricultural Research Stations (ZARS) to take up the additional functions of KVK, and Strengthening of Directorate of Extension of State

Agricultural Universities and Zonal Units have been recently undertaken under the National Agricultural Technology Project (NATP).

The Council aids, promotes and coordinates agricultural education programmes at the national level. The educational programmes are carried out through its network of universities. In addition to these, 4 National Institutes of the Council, *viz.*, Indian Agricultural Research Institute (IARI), Indian Veterinary Research Institute (IVRI), National Dairy Research Institute (NDRI), and Central Institute of Fisheries Education (CIFE) have been granted the status of deemed university. These institutes offer post-graduate degree programmes besides undertaking research in their respective areas.

Achievements

The agriculture research system has significantly contributed to productivity growth in almost all the sectors of agriculture. For instance with reference to 1950, the productivity gains are nearly 3.3 times in foodgrains, 1.6 times in fruits, 2.1 times in vegetables, 5.6 times in fish (aquaculture), 1.8 times in milk and 6.4 times in eggs. Development of about 3200 high yielding improved varieties of different crops and their production technologies and preservation of 2.2 lakh germplasm accessions of agri-horticultural crops and their wild relatives for future use in breeding programmes, are some remarkable achievements of the National Agricultural Research System. It is also revealing to note that the modern technologies have contributed to saving of the area, other resources and in increasing revenues and exports. For instance, modern varieties of rice saved nearly 39 million hectare and the wheat varieties 37 million hectare area.

Research in livestock and poultry sector has provided more options for increased income generation. The fisheries sector is also growing at an incremental pace, with an annual production of 5.9 million tonnes. Along with the livestock and poultry sector, it also contributes significantly to domestic nutritional security and rural development. Research for development of farming implements has helped to make agricultural operations modern and less labour-intensive apart from saving the operational cost.

The programme on frontline demonstrations of improved production technologies, on-farm demonstrations of integrated nutrient and pest management, and a variety of training programmes involving farmers have shown encouraging results by successfully demonstrating the potential of new and improved technologies.

Keeping India Vision 2020 in view, each ICAR institution has formulated its Perspective Plan document, 2020. Adopting a bottom up approach, ICAR has also formulated its Vision 2020 document which has laid a broad road map for the next 20 years. ICAR scientists and peers have strived hard to set up a demand-driven research and development agenda. The gains from green revolution were not only consolidated but also several new grounds were broken for attaining new gains. Technological improvements were made in several non-traditional crops like soyabean, new crop seasons were explored in rabi maize, rabi sorghum, summer groundnut and summer pearl millet. Hybrid breeding approach was extended to rice, safflower, castor, pigeon pea and mustard.

Increased yields and stability were achieved through shifting away from kharif season production to rabi season production. Diversification covering fruits and vegetables, flowers, animal enterprises, poultry, fish and agro-forestry are also emphasised aiming at golden revolution. All these efforts added to defending the gains of green revolution, extending the gains to new areas and making new gains in several important areas contributing to accelerated agricultural transformation.

GROWING DEMAND

Notwithstanding the phenomenal success in agriculture, increasing population and per capita income demand higher production and productivity levels. The annual increase in demand by the end of the 10th Five Year Plan will be 2.6 million tonnes in rice, 2.2 million tonnes in wheat, 1.6 million tonnes in pulses, 4.2 million tonnes in fruits, 2.5 million tonnes in vegetables, 7.8 million tonnes in milk and 0.6 million tones in fish. The 10th Plan fixes a target growth rate of 4 per cent in agriculture which in itself is challenging, keeping in view the achieved growth rate of 1.95 per cent in food production during the IX Plan.

After attaining self-sufficiency in food, concerns on increasing factor productivity, improving resource efficiency and economic access to food combined with better nutrition, health and shelter became important which were addressed through diversification of agriculture. At this point of time, natural resource management problems became critical, leading to plateauing of yield and, therefore, conservation of natural resources such as soil, water and genetic wealth. The concerns of global competitiveness and system's sustainability became critical around 2000 which are now receiving priority attention.

The eight-fold current concerns of the ICAR include –input use efficiency and consequent reduction in cost of cultivation; diversification of agriculture while conserving natural resource base; rainfed agriculture; post-harvest management; sustainability, equity and gender issues; motivated and creative human resource; easy access and quick disposal of information and building a culture of trust and sense of belonging among scientists, farmers, extension workers, young entrepreneurs and NGOs. Thus, the concerns are evolving and the system is responding well. The contributions of technology in particular have been unique.

TECHNOLOGY LED-PROSPECTIVE AGRICULTURAL GROWTH

Indian agriculture is passing through many challenges following increasing demands of foodgrains, growing competition among nations for export of value-added quality produce, rising cost of production and dwindling sources of water. But they offer new opportunities for quality improvement, value-added products development, processing and enhancing the shelf life of agri-products through which India can compete with the rest of the world. New technologies like biotechnology, hybrid technology and bio-informatics will have their imprint on our success for meeting the challenges of food security and sustainability.

The six new initiatives of the Department of Agricultural Research and Education during the Xth Plan, are – Network Project on organic farming, Impact of Climate Change on Agricultural Productivity, Networking on Transgenics, National Bureau of Indian Veterinary Type Culture, Network on Biosystematics of Insects and National Research Centre for Agricultural Extension. Rainbow revolution encompassing an all-round growth in the production of foodgrains, edible oils, fruits, vegetables and animal and fish products and evergreen revolution in agriculture through diversified agriculture, precision farming, resource conservation and value addition to provide in real sense, the required food, nutrition and environmental security are the hopes of the future.

2

Agriculture and Rural Development

AGRICULTURE IN INDIA

Agriculture in India is one of the most prominent sectors in its economy. Agriculture and allied sectors like forestry, logging and fishing accounted for 18.6% of the GDP in 2005 and employed 60% of the country's population. It accounts for 8.56% of India's exports. About 43% of India's geographical area is used for agricultural activity.

Despite a steady decline of its share in the GDP, agriculture is still the largest economic sector and plays a significant role in the overall socio-economic development of India. The monsoons play a critical role in the Indian sub-continent's agriculture in determining whether the harvest will be bountiful, average, or poor in any given year. The entire rainfall in the sub-continent is concentrated in the few monsoon months.

Agriculture in India is constitutionally the responsibility of the states rather than the central government. The central government's role is in formulating policy and providing financial resources for agriculture.

The government administers prices of essential commodities to protect farmer's interests. It also administers other commodities which are produced by government-controlled companies, like petroleum, coal, nitrogenous fertilizers, etc. Other than these, most agricultural commodity markets operate under the normal forces of demand and supply.

HISTORY OF INDIAN AGRICULTURE

Evidence of the presence of wheat and some legumes in the 6th millennium BC have been found in the Indus Valley. Oranges were cultivated in the same millennium. The crops grown in the valley around 4000 BC were typically wheat, peas, sesame seed, barley, dates and mangoes. By 3500

BC cotton growing and cotton textiles were quite advanced in the valley.

By 3000 BC farming of rice had started. Other monsoon crops of importance of the time was cane sugar. By 2500 BC, rice was an important component of the staple diet in Mohenjodaro near the Arabian Sea. The Indus Plain had rich alluvial deposits which came down the Indus River in annual floods. This helped sustain farming that formed basis of the Indus Valley Civilization at Harappa. The people built dams and drainage systems for the crops. By 2000 BC tea, bananas and apples were being cultivated in India. There was coconut trade with East Africa in 200 AD. By 500 AD, egg plants were being cultivated.

Agricultural Movements

Green Revolution

After independence, considering India's growing population, the government took steps to increase the food production. Yields per unit area of all crops have grown since 1950. The 1970s saw a huge increase in India's wheat production. This is known as the Indian Green Revolution. Reasons for the growth are the special emphasis placed on agriculture and steady improvements in irrigation, technology, application of modern agricultural practices and provision of agricultural credit and subsidies

Operation Flood

Operation Flood was the name of a rural development programme started by the National Dairy Development Board (NDDB) in 1970 with the objective of creating a nationwide milk grid. This movement followed the Indian green revolution and helped in alleviating poverty and famine levels from dangerous proportions in India during the era. It resulted in India becoming the largest producer of milk and milk products, so it is also called the White Revolution of India.

Production

India is ranked second in the world in terms of agricultural output. India is the largest producer in the world of milk, cashew nuts, coconuts, tea, ginger, turmeric and black pepper. It also has the world's largest cattle population (193 million). India ranks second worldwide in farm output.

It is the second largest producer of wheat, rice, sugar, groundnut and inland fish. It is the third largest producer of tobacco. India accounts for 10 per cent of the world fruit production with first rank in the production of banana and sapota. Given below is a chart of trend of output of cereals and major foodgrains as published by the Department of Food and Public Distribution with figures in tonnes.

AGRICULTURE, GROWTH AND POVERTY REDUCTION

History shows that different rates of poverty reduction over the past 40 years have been closely related to differences in agricultural performance – particularly the rate of growth of agricultural productivity.

In Asia, the rapid productivity gains of the Green Revolution increased producers' incomes, raised labourers' wages and lowered the price of food. In addition, new livelihood opportunities were generated when success in agriculture provided the basis for economic diversification. However, despite decades of investment in new agricultural technology and rural development, hunger and poverty continue to plague large areas of the developing world. The problem is particularly acute in sub-Saharan Africa, where progress towards the Millennium Development Goals is slowest.

While increasing agricultural productivity perhaps remains the single most important determinant of economic growth and poverty reduction, serious doubts are emerging as to whether agricultural productivity can be further increased, where it is most needed, and what part, if any, small-scale farming will play in the future. Development experts need greater understanding of the links between agricultural productivity and poverty. They also need to assess just how far they have changed and the extent to which small-scale agriculture can remain a ladder out of poverty for millions of poor people living in rural areas. Links between agriculture and poverty reduction are forged through four 'transmission mechanisms':

- direct and relatively immediate impact of improved agricultural performance on rural incomes;
- impact of cheaper food for both urban and rural poor;
- agriculture's contribution to growth and the generation of economic opportunity in the non-farm sector; and
- agriculture's fundamental role in stimulating and sustaining economic transition, as countries (and poor people's livelihoods) shift away from being primarily agricultural towards a broader base of manufacturing and services.

However, the potential for future poverty reduction through these transmission mechanisms depends on the extent to which agricultural productivity can be increased where it is most needed or, more simply and directly, the extent to which the Green Revolution can be replicated in Africa. Whether and how this might be achieved has been the subject of protracted debate amongst researchers and policy makers. The main questions in this debate are:

- Where to focus development efforts – high potential areas where development options are greater or poorer areas where the potential and options are less but the incidence of poverty is greatest?

- Who to focus on – the poorest smallholders who produce for subsistence and have limited engagement with markets, or larger-scale farmers whose success can be an engine of growth and generate wealth and jobs for rural households?
- What to do in less favoured areas – should agriculture be promoted amongst poor households in areas of low agricultural potential that are remote from markets and inputs supplies? Or should the focus be on employment generation to enable poor households to buy in food?
- What role for technology – should primacy in development efforts be given to yield-raising technology or to less intensive approaches that minimise variation?
- Which crops – should the emphasis be on crops that will be largely consumed within the households, or on income-generating cash crops?

Contrasting answers to these questions produce different policy positions. Those who are optimistic about the role of smallholders emphasise the need for increased direct investment in agriculture and rural development, with support focusing on creating institutions that will encourage and support smallholder-led agricultural development. They believe it is important to improve the productivity of staple food crops that are not traded internationally, but consumed by the poor and traded locally.

In contrast, smallholder pessimists emphasise the need to achieve the best outcome possible from rapidly changing global markets for agricultural produce. They believe that commercial production (probably on larger farms) of non-staple cash crops should be encouraged, particularly those that result in robust links to the non-farm sector where the main source of employment for the rural poor would be found. Influencing international policy processes to ensure access to developed country markets is a priority for this position, as is improving the human capital assets of the rural poor so they can take up opportunities in growth areas as they occur.

What is the Issue?

Poverty has fallen rapidly over the past 40 years, but at different rates around the world. Asia has achieved the most rapid poverty reduction, particularly China, but also India and South East Asia. In contrast, little if any progress has been made in sub-Saharan Africa, where the number of people living on less than one dollar a day – the internationally agreed definition of absolute poverty – has doubled over the past 20 years (World Bank, 2004a).

Historically, rates of poverty reduction have been very closely related to agricultural performance – particularly to the rate of growth of agricultural productivity. In simple terms, this indicates that the countries that have increased their agricultural productivity the most have also achieved the greatest reductions in poverty.

In Asia rapid productivity gains, achieved largely through the technological advances of the Green Revolution, provided a 'fast-track' route out of poverty by directly increasing producers' incomes and labourers' wages, by lowering the price of food and by generating new livelihood opportunities as success in agriculture provided the basis for economic diversification. However, despite decades of investment in new agricultural technology and rural development, hunger and poverty continue to plague large areas of the developing world.

The problem is particularly acute where people depend on rain-fed agriculture, in particular sub-Saharan Africa, where the impact of new technologies has been less apparent and agricultural productivity has generally stagnated and even fallen in some areas.

Achieving the Millennium Development Goal (MDG) of halving the proportion of people living in absolute poverty by 2015 will depend largely on increasing agricultural productivity, which remains perhaps the single most important determinant of economic growth and poverty reduction. This fact is not lost on developing countries or their development agency partners, who are seeking ways to stimulate agricultural development. But serious doubts are emerging as to whether agricultural productivity can be increased where it is needed most, and what part, if any, small-scale farming will play in the future. Underlying such doubts is a concern that the context in which small-scale agriculture could achieve productivity gains today is very different to that which prevailed in Asia during the halcyon days of the Green Revolution. Inherent differences in production capabilities and quite fundamental changes in the international agricultural context are combining to create a set of circumstances that are probably far less conducive to achieving the type of productivity transformation witnessed in Asia.

Development experts need greater understanding of the links between agricultural productivity and poverty. They also need to assess just how far they have changed and the extent to which small-scale agriculture can remain a ladder out of poverty for millions of poor people living in rural areas.

A SAGA OF SUCCESS

From a nation dependent on food imports to feed its population, India today is not only self-sufficient in grain production but also has a substantial reserve. The progress made by agriculture in the last four decades has been one of the biggest success stories of free India. Agriculture and allied activities constitute the single largest contributor to the Gross Domestic Product, almost 33 percent of it. Agriculture is the means of livelihood of about two-thirds of the workforce in the country.

This increase in agricultural production has been brought about by bringing additional area under cultivation, extension of irrigation facilities, the use of improved high-yielding variety of seeds, better techniques evolved through agricultural research, water management, and plant protection through judicious use of fertilisers, pesticides and cropping practices.

Crops

The 1970s saw a multi-fold increase in wheat production that heralded the Green Revolution. In the next decade rice production rose significantly; in 1995-96, rice production was 79.6 million tonnes. The total grain production crossed 211 million tonnes in 2001-02, a big leap from 51 million tonnes in 1950-51. To carry improved technologies to farmers, a National Pulse Development Programme, covering 13 states, was launched in 1986. Efforts to boost pulse production were augmented further by the Special Food Production Programme. In 2001-02, pulse production was 13.52 million tonnes. With some states offering more than the statutory minimum price, sugar cane production also received a boost, and in 2001-02 a record 292.2 million tonnes was registered.

Irrigation

As efforts continued to increase the irrigation potential in the country, the last 40 years saw the gross irrigated area reach 85 million hectares. Flood forecasting has become an important activity over the years. Over 500 hydrological stations collect and transmit data through 400 wireless stations for issuing forecasts for 157 sites. About 5,000 forecasts are issued in a year with 94 percent accuracy. The country also receives international support, with the World Bank as a primary source, for developing its water resources. International cooperation is also envisaged in setting up a National Centre for Information on Water and Power. As there is a broad seismic belt in the country, particularly along the Himalayan, and the Kutch region and parts of Maharashtra, a scheme is being evolved to collect all data on seismic activity at various dam sites.

Fertilisers

The fertiliser industry in India has grown tremendously in the last 30 years. The Government is keen to see that fertiliser reaches the farmers in the remote and hilly areas. It has been decided to decontrol the prices, distribution and movement of phosphatic and potassic fertilisers. Steps have been taken to ensure an increase in the supply of non-chemical fertilisers at reasonable prices. There are 66 fertiliser quality control laboratories in the country. Since biofertilisers are regarded as an effective, cheap and renewable supplement to chemical fertilisers, the Government is implementing a National Project on Development and Use of Bio-fertilisers. Under this scheme, one national and six regional centres for organising training, demonstrating programmes and quality testing of bio-fertilisers has been taken up. It was a challenging decision of the Government to take Bombay High gas through a 1,700 km pipeline to feed fertiliser plants located in the consumption centres of North India.

However, the major policy which has ensured the growth of the fertiliser industry is the thrust on accelerating fertiliser consumption by fixing, on the one hand, low and uniform price for fertilisers, and on the other hand providing the manufacturers adequate compensation through the retention price and subsidy scheme. As expected, fertiliser nutrient demand has gone up from 0.29 million tonnes in 1960-61 to 16.7 million tonnes at the end of 2000-01, compared to 12.15 million tonnes during 1992-93.

Fisheries

Fish production achieved an all-time high of 5.6 million tonnes at the end of 2001-02. Programmes that have helped boost production include the National Programme of Developing Fish Seeds, Fish Farmers' Development Agencies and Brackish Water Fish Farmers' Development Agencies. The Central Institute of Fisheries Nautical and Engineering Training trains the necessary manpower. To diversify fishing methods and introduce processed fish products on a semi-commercial scale, an Integrated Fisheries Project has been launched. A National Fisheries Advisory Board has also been established.

FOOD PROCESSING

The Ministry of Food Processing Industries, set up in July 1988, is the central agency of the Government responsible for developing a strong and vibrant food-processing sector with a view to create increased job opportunities in rural areas, enable the farmers to reap benefit from modern technology, create surplus for exports and stimulating demand for processed food. A new seeds policy has been adopted to provide access to high-quality seeds and plant material for vegetables, fruit, flowers, oilseeds and pulses, without in any way compromising quarantine conditions. Initiatives have been taken to encourage private sector investment in the food processing industry.

AGRICULTURAL RESEARCH

The apex body for education, research, training and transfer of technology in the field of agriculture is the Indian Council of Agricultural Research (ICAR), established in 1929. India's transformation from a food deficit to a food surplus country is largely due to ICAR's smooth and rapid transfer of farm technology from the laboratory to the land.

ICAR discharges its responsibilities through 43 research institutes, four national research bureaux, 20 national research centres, nine project directorates, 70 all-India coordinated research projects, and 109 Krishi Vigyan Kendras (farm science centres). Besides, the programme of Agricultural Education is coordinated by ICAR with the curricula and other normative guidance given to the 26 agricultural universities and four national research institutes.

OILSEEDS PRODUCTION

A Technology Mission on Oilseeds was launched in 1986 to increase production of oilseeds in the country and attain self-sufficiency. Pulses were brought under the Technology Mission in 1990. After the setting up of the Technology Mission, there has been consistent improvement in the production of oilseeds. The oilseed production, which was 108.3 lakh tonnes in 1985-86, has increased to 247.30 lakh tonnes in 1998-99, the highest production so far. However, due to unfavourable weather conditions prevailing in the major oilseeds-growing states, the oilseed production during 1999-2000 and 2000-01 suffered some decline.

The increase in production was largely contributed by soyabean, rapeseed and mustard. Production of pulses has seen many ups and downs, which is expected to be checked under the mission. The country grows mainly nine oilseeds, with groundnut, rapeseed and mustard accounting for 62 percent of total production. Lately, soyabean and sunflower have shown major growth potential.

DRINKING WATER

A Technology Mission on Drinking Water and Related Water Management has been constituted to cover the residual problem villages and provide potable water at 40 litres per capita per day, and 70 litres per capita per day in desert areas inclusive of 30 litres for cattle. The mission is tackling the problem through 55 mini-missions in project districts and countrywide problem-oriented sub-missions. A Village Level Operation and Maintenance (VLOM) pump called India Mark-II has been developed and is being exported to 40 countries.

INSTABILITY ASSOCIATED WITH IMPROVED AGRICULTURAL TECHNOLOGY

There has been much concern over the possibility that improved crop varieties are associated with increased output instability in addition to higher yields. If this is the case, it could deter adoption of more productive technology and hence retard agricultural development.

Instability at the Farm Level

In discussions of farmers' adoption of HYVs, it is often assumed that new seeds have more volatile yields. Indigenous landraces often have the characteristic of being fairly unresponsive to improved management and increased inputs, but also robust in the face of unfavorable agroclimatic conditions, such as drought. HYVs, on the other hand, often are characterized as being highly responsive to management and inputs, but very low yielding in the event of bad weather. If this is the case, then plant breeders must breed on the basis of one strategy (high average yield) for high potential areas and another (risk minimization) for low potential areas.

Research on pearl millet, which is commonly grown in unfavorable agroclimatic areas, suggests that the above characterization of traditional varieties and HYVs is not necessarily correct. Witcombe (1989), in a study of Pakistan and India, found that good performance of a particular millet seed over all environments appears to indicate good performance in environments of low potential. In three years out of his four year study, the highest yielding entry across all environments was also one of the two highest yielding entries in the lowest yielding environment. These results suggest that the typical plant breeder's strategy of selecting among the highest yielding seeds across all environments is satisfactory. Farmers who adopt HYVs do not necessarily subject themselves to greater risk of catastrophic loss in the event of bad weather.

Farmers' Strategies to Reduce and Cope with Risk. Walker and Jodha (1986) point out that dryland farmers have various methods to reduce their exposure to crop production risk. Cultural practices play an important risk-reducing role; they include planting different crops with relatively low covariate yield (either in an intercrop or on separate fields); diversifying spatially by operating multiple plots with different environmental characteristics; and staggering planting dates in the face of variable rainfall patterns. Sharecropping is a common tenancy arrangement that distributes risk between the tenant and landlord. Many farmers have multiple sources of income, reducing risk if they have low covariation.

Farmers also have various mechanisms to cope with risk that they cannot eliminate. For example, they can borrow from local stores or money lenders, draw down food stocks or savings, sell assets, obtain transfers from relatives, participate in government relief programs, or migrate. Most of these options are not particularly desirable; for especially poor people they can be quite devastating: selling assets or going into debt may make a family permanently worse off even after drought is over. Jodha (198_) indicates that some families will reduce their food consumption as much as possible before parting with their assets; this has obvious negative short term health implications that are particularly severe for those who consume only minimum requirements to begin with.

Effect of Risk on Technology Adoption. While weather-related risk undoubtedly presents great hardship for a very large number of people, Walker (1989b) indicated that it may not be as important as generally believed in adoption of new technologies in rainfed areas.

The tradeoff between expected income and the variance of that income suggests that given farmers' measured risk preferences, the overall effect on adoption decisions is modest. Also, Walker stressed that yield instability does not translate into major variability in income due to farmers' mechanisms to absorb risk. As mentioned above, multiple sources of income and diverse cropping patterns mean that yield variability of one crop only affects a portion of the income from crops. Equally important, in many cases farmers can adjust the area under each crop depending on the weather at the start of the season. For example, dryland rabi (postrainy season) sorghum farmers in black soil areas know at the start of the season how much moisture is available and adjust their cropped area accordingly. Similarly, castor farmers in Andhra Pradesh know that pest attacks are more prevalent when the rainy season begins late, so they plant less castor. And farmers with irrigation from tanks or wells know roughly how much water will be available in the postrainy season, so they adjust planted area accordingly. Hazell and Ramasamy (1991), for example, found that sharply reduced paddy production in North Arcot, Tamil Nadu, in the drought year 1982-83 resulted from a fall in area, not yield. Not surprisingly, Walker (1989b) also found that variations in cultivated area This is why SWC programs are sometimes referred to as "drought-proofing" measures.

exceeded those in yield. He also found that given farmers' diversified farming and livelihood strategies, the large variations in yield translated into only small variations in income. The variations were so small that even if plant breeders could develop varieties with perfect stability (zero variation), the contribution to household income stability would amount to less than 1% for most of the crops studied, and a maximum of 2.9% for paddy (which, paradoxically, had the least yield variability to begin with).

Walker's (1989b) results demonstrate the importance of looking beyond yield variability of a single crop to variations in all the crops in a given household, village or region, and beyond crop variation to income variation. From this perspective, yield instability does not appear to be a major determinant of adoption, and hence not a top priority for plant breeders. He pointed out (in his 1989a study) that policies related to international trade and storage between surplus and deficit years can be more cost-effective in coping with increasing yield instability.

Soil and water conservation (SWC) investments also are associated with a variety of risks. First, erosion itself is a matter of risk. For example, some plots may be at risk of productivity loss from continuous, gradual erosion, but others may be more susceptible to significant erosion only in the event of a once-in-five-years or once-in-fifty-years storm. For such lands, in normal years there is no gain from investments to reduce erosion, but in exceptional years the gain — actually the avoided loss — may be quite high. Second, often SWC practices serve to conserve soil moisture as well as reduce erosion. Soil moisture retention is more likely to offer immediate, productivity-increasing benefits than erosion prevention. However, these benefits may exceed the costs only when rainfall is unusually low or unevenly distributed, so that moisture stress constrains productivity.12 In a good rainfall year, on the other hand, short term gross returns to increased moisture retention may be low or zero, and in a very high rainfall year they may even be negative if they lead to waterlogging. Little information is available regarding the impact of risk on Indian farmers' SWC investments.

Instability at the Aggregate Level

Ramakrishna (1993) used a log function to calculate CVs to compare instability of production, yield and area cultivated for cereals between the pre-green revolution and green revolution periods. Ramakrishna found that for food grains, output was more stable during the green revolution than before. Yields were about equally stable between the two periods, and area became more stable. For other crops, on the other hand, Ramakrishna found that output and yield instability both increased during the green revolution, while area again became more stable. Combining food grains and other crops, output and yield instability increased slightly during the green revolution, while area instability was constant. On the whole, Ramakrishna's data suggest that the green revolution had little impact on agricultural instability.

Hanumantha Rao (1994) took a slightly different approach and found different results. Instead of taking CVs of output levels, he took the standard deviation of annual output growth rates as a measure of instability. He found that the standard deviation of output growth of food grains was 8.1 in 1950-51 but 11.4 between 1968-85, and took this to indicate that output instability increased with the green revolution. He also found that the standard deviation rose from 9.4 in the first decade of the green revolution to 11.8 in the second decade. Hanumantha Rao did not provide detailed figures regarding yield and area instability. His figures do not really offer a good comparison to those of Ramakrishna, because his pre-green revolution figures are based only on a single year. Also, it is difficult to compare the magnitude of Hanumantha Rao's figures (expressed as the standard deviation of output growth rates) to Ramakrishna's (expressed as the CV of the level of output). Since we do not know the growth rates from which Hanumantha Rao calculated the standard deviation, we cannot relate his standard deviations to Ramakrishna's CVs. Hazell (1982) examined cereal crop production between 1954-55 and 1964-65 (before the green revolution) and 1967-68 and 1977-78 (during the green revolution, after the introduction of high yielding varieties).13 He found that the CV of production increased by about 50%, from 0.04 to 0.059 between the two periods. Hazell hypothesized that if the increased instability were due to HYVs, variances in production within states would have to rise. However, he found that changes in yield covariances were much more important than changes in yield variances. Only about 18 percent of the increase in variance of total cereal production resulted from changes in crop production variances; the remaining 82 percent was explained by changes in covariances; interstate covariances within crops contributed 41 percent to the change in variance in total cereal production. As a result, Hazell concluded that HYVs were probably not the primary cause of increased variability. In a later paper, Hazell (1984) suggested that HYVs could possibly affect yield covariances of maize in India and the United States, because the narrower genetic base of improved varieties would make them susceptible to common yield inhibitors such as pests and diseases. This would then be a contributing factor to greater variability in national ceral production.

Walker (1989a) studied changes in yield and output variability of sorghum and pearl millet in India resulting from the spread of HYVs. This study followed Hazell's 1984 study relating the rising covariance of yields between states to the spread of HYVs; it used districtlevel data for the 48 largest sorghum districts and the 40 largest pearl millet districts. Walker found that variability had indeed increased significantly between the period 1956-57 to 1967- 68 and 1968-69 to 1979-80, from 8% to 16% for sorghum and 11% to 34% for pearl millet. He also found that covariance of yields across districts was by far the most important factor in overall output variance of sorghum and pearl millet. Walker examined several possible causes of increased covariance, including 1) changes in rainfall covariance, 2) changes in irrigated area, and 3) adoption of HYVs. He found evidence that provided weak support for each of these possible sources of covariance. However, he also stressed that the contribution of HYVs to increased instability is dwarfed by their contribution to productivity, so he did not recommend changes in existing breeding strategies that develop HYVs for adoption over a large area.

Walker's (1989a) study helps clarify that numerous factors can contribute to changes in stability. Researchers who find that stability increased or decreased with the green revolution should hesitate before proclaiming that HYVs were the cause. Changes in cropping patterns, fluctuations in weather, and the quality of land on which a particular crop tends to be planted all can have an impact on the variations in yield. More importantly, variations in area may either counteract or reinforce those in yield; it is important to know the composition of output fluctuations between yield and area variations. The studies by Hazell and Walker also highlight the need to distinguish between variability at the regional or national levels from farm or field level risks. Because of the dominance of covariance relations in aggregate production data, variability can increase at the aggregate level even while farm level variability changes little or not at all. Although diets were stable, health problems increased due to a shortage of clean drinking water in the village resulting from the drought.

DROUGHT RISK

The preceding discussion on output instability has underemphasized the risks associated with the specific problem of severe, prolonged drought. When severe drought occurs over a wide area, yield and income risks can be particularly great. In Hazell and Ramasamy's (1991) study of North Arcot district, Tamil Nadu, for example, they found that the 50% decline in paddy production in their study villages was replicated throughout the district and beyond. Incomes fell by 50% on average as even nonfarm incomes fell sharply. Consumption expenditure fell by 50% on average, and diet quality deteriorated. Clearly, the effects of drought were covariate across villages and sectors. This problem of covariate risk is particularly challenging because poor performance of one income source, or one location, cannot necessarily be compensated by better performance in another. As a result, people may face severe hardships in the event of drought.

In the Bidinger et al (1990) study, on the other hand, despite a similar 50% decline in income, dietary intake did not change compared to the pre-drought situation.14 People maintained their dietary intake through increases in temporary migration and consumption credit, and government rice subsidies that kept prices low and stable, enabling people to translate meager incomes into normal diets. The number of people who migrated only increased by about 10%, but they stayed away much longer on average. Interest rates did not show the expected increase as demand for consumption credit rose, even though the bulk of borrowing was done through the informal village credit market rather than the formal banking system, which contributed little. Distress sales of land and livestock were rare, but two families did sell all their assets and move permanently to become urban laborers.

GOVERNMENT INTERVENTIONS TO MANAGE RISK

Severe droughts impact negatively on most rural households simultaneously and are therefore difficult to manage through traditional risk sharing and coping strategies. As a result, government policy may sometimes play an important role in helping farmers manage risk. While some government interventions are well-established, other ideas remain relatively untested. Likewise, even where farmers and rural communities have developed effective mechanisms to manage production risk, it is not clear what the costs of these mechanisms are in terms of reduced production efficiency. Perhaps market-based or government-sponsored alternatives can be introduced that protect farmers from the effects of risk but without requiring diverse income sources and fragmented agricultural holdings. Anderson and Hazell (1994) point out that more information is needed about the costs of these risk-reducing and coping mechanisms. We briefly review government measures to manage risk and discuss some additional possible government approaches.

Most of the existing drought management efforts. These are poverty allevation programs to cope with the effects of drought, namely employment programs and food subsidies. There is no need to repeat the discussion of these programs here. Bidinger et al (1990), stress the importance of food subsidies in mitigating the effects of drought in their study, but they also point out that a timely public works program could have prevented much of the unemployment and debt experienced by laborers in the village.

Crop insurance is provided by the public sector in many countries. The impetus for such programs often originates in governmental concern about catastrophic risks such as drought, or the desire to reduce the incidence of loan defaults to banks. With few exceptions, the financial performance of public crop insurers has been ruinous (Hazell 1992). To be financially viable without government subsideis, an insurer needs to keep the average value of its annual outgoings—indemnities plus administration costs—below the average value of the premiums it collects from farmers. In practice, many of the larger insurance programs pay out $2.50 or more for every dollar of premium they collect from farmers. The difference is paid by governments, at costs varying from $10 to $400 per insured hectare. Even at these levels of subsidy, many farmers are still reluctant to purchase insurance. As such, many crop insurance programs are compulsory, either for all farmers growing specified crops or for those who borrow from agricultural banks. The primary reason for the high cost of public crop insurance schemes is that they invariably attempt to insure risks that are prone to severe moral hazard problems, whereby farmers have incentive to lie about their production levels, or to allow their crops to fail in order to receive insurance payments (Hazell 1995a). These risks include many climate, disease and pest risks that are difficult to quantify and assess, and whose damage can be influenced by armers' management practices. The problem is aggravated by a common practice of insuring "target" yields rather than compensating for actual losses. But this is not the only reason for failure.

Another overwhelming factor is the incentive problem that arises oce the government establishes a pattern of guaranteeing the financial viability of an insurer. If the insurance staff know that any losses will automatically be covered by governmetn, they have little incentive to pursue sound insurance practices when setting premiums and assessing losses. In fact, they may find it profitable to collude with farmers in filing exaggerated or falsified claims. Yet another common reason for failure has been that governments undermine public insurers for political reasons. In Mexico, the total indemnities paid has borne a strong statistical relationship with the electoral cycle, increasing sharply immediately before and during election years, and falling off thereafter. In the USA, the government has repeately undermined the national crop insurer (FCIC) by providing direct assistance to producers in disaster areas. Why should farmers purchase crop insurance against major calamities (including drought) if they know that farm lobbies can usually apply the necessary political pressure to obtain direct assistance for them in times of need at no financial cost? Another reason for their high cost is that crop insurers tend to be too specialized, focusing on specific crops, regions and types of farmers, particularly when the insurance is tied to credit programs designed to serve particular target groups identified by the government. Without a well-diversified insurance portfolio, crop insurers are susceptible to covariability problems, a face the prospe t of sizable losss in some years. Since public insurers are rarely able to obtain commercial reinsurance or contingent loan arrangements, thsi specialization increasestheir dependence on the government. Crop insurances is provided by the public sector in many countries. The impetus for such programs often originates in governmental concern about catastrophic risks such as drought, or the desire to reduce the incidence of loan defaults to banks. With few exceptions, the financial performance of public crop insurers has been ruinous (Hazell 1992). To be financially viable without government subsidies, an insurer needs to keep the average value of its annual outgoings — indemnities plus administration costs — below the average value of the premiums it collects from farmers. In practice, many of the larger insurance programs pay out $2.50 or more for every dollar of premium they collect from farmers. The difference is paid by governments, at costs varying from $10 to $400 per insured hectare. Even at these levels of subsidy, many farmers are still reluctant to purchase insurance. As such, many crop-insurance programs are compulsory, either for all farmers growing specified crops, or for those who borrow from agricultural banks. The primary reason for the high cost of public crop-insurance schemes is that they invariably attempt to insure risks that are prone to severe moral hazard problems whereby farmers have incentive to lie about their production levels, or to allow their crops to fail in order to receive insurance payments (Hazell 1995a). These risks include many climate, disease and pest risks that are difficult to quantify and assess, and whose damage can be influenced by farmers' management practices. The problem is aggravated by a common practice of insuring "target" yields rather than compensating for actual losses. But this is not the only reason for failure.

Another overwhelming factor is the incentive problem that arises once the government establishes a pattern of guaranteeing the financial viability of an insurer. If the insurance staff know that any losses will automatically be covered by government, they have little incentive to pursue sound insurance practices when setting premiums and assessing losses. In fact, they may find it profitable to collude with farmers in filing exaggerated or falsified claims.

Yet another common reason for failure has been that governments undermine public insurers for political reasons. In Mexico, the total indemnities paid has borne a strong statistical relationship with the electoral cycle, increasing sharply immediately before and during election years, and falling off thereafter. In the USA, the government has repeatedly undermined the national crop insurer (FCIC) by providing direct assistance to producers in disaster areas. Why should farmers purchase crop insurance against major calamities (including drought) if they know that farm lobbies can usually apply the necessary political pressure to obtain direct assistance for them in times of need at no financial cost? Another reason for their high cost is that crop insurers tend to be too specialized, focusing on specific crops, regions and types of farmers, particularly when the insurance is tied to credit programs designed to serve particular target groups identified by the government. Without a well-diversified insurance portfolio, crop insurers are susceptible to covariability problems, and face the prospect of sizable losses in some years. Since public insurers are rarely able to obtain commercial reinsurance or contingent loan arrangements, this specialization increases their dependence on the government. Public crop insurers also tend to have high administration costs. This is partly because they often insure small-scale farmers, but also because crop-insurance work is very seasonal, and the absence of a well-diversified portfolio means that staff and field equipment are underemployed for significant parts of the year.

There is no convincing evidence that public subsidization of crop insurance has been socially beneficial. Indeed, social benefit-cost analyses of the Mexican and Japanese schemes show negligible social returns in relation to their high costs (Bassoco et al. 1986, Tsujii 1986). Nor is there much evidence that it has increased agricultural lending or benefited agricultural banks. In a rare study, Pomareda (1984) compared the performance of insured and uninsured loans in the portfolio of the Agricultural Development Bank (BDA) of Panama. Insured loans had slightly higher and more stable returns than uninsured loans. They were also repaid and cleared from the books closer to their expected duration. But the overall gains to the Bank were modest, and could have been achieved more easily at no cost to the government simply by allowing a 2 percent increase in the interest rate that BDA charged its borrowers. This would also have been cheaper for the borrowers than the premium rates they paid for the compulsory insurance.

In order to overcome the major problems associated with crop insurance, and to substantially reduce its administration costs, several authors have proposed area-based yield insurance (e.g. Halcrow, 1948; Dandekar, 1977; Miranda, 1991). Under this proposal, the crop yield for a *homogeneous region* is insured, and all insured farmers in the region pay the same premium and receive the same indemnity. Indemnities are paid whenever the average yield for the region falls below some critical level irrespective of the actual yields obtained by individual farmers. Premiums are calculated on the basis of year-to year variations in the average yield for the region, and would vary from one homogeneous region to another in accordance with differences in risk levels. This approach reduces moral hazard problems, and hence broadens the range of yield risks that can be viably insured. Moreover, since the premiums and indemnities are identical for all insured farmers in a region, it avoids the *adverse selection* problem. The latter refers to the situation in which farmers facing below-average risk tend to drop out of insurance programmes if they are charged premium rates based on average risk levels but are paid indemnities based on their own losses. Also, by eliminating the need for field inspections and loss assessments, the cost of administering an area-based scheme could be kept very low. Providing farmers pay their premium, it is not really necessary that they even grow the crop that they have insured.

Despite its appeal and its potential scope for reaching small-scale farmers, there are problems with the proposal. First, the insurance will be attractive to individual farmers only if their yields are highly correlated with the average yield for the region. Dandekar (1977) argues the homogenous areas can be defined in India in which inter-farm yield correlations are positive, but their is a growing body of micro-evidence showing that yield correlations between plots within the same village, or even within the same farm, are surprisingly low (e.g. Walker and Jodha, 1986). Small differences in ground contours, slope, and wind and sun exposure can lead to substantial differences in the yield damage caused by unfavorable climatic, pets and disease events, as can a few days difference in planting dates or the crop varieties grown.

Second, the scheme is subject to severe covariability problems. When the average yield is below the critical value in a region, all the insured farmers have to be compensated simultaneously. An unsubsidized insurer could hope to survive only if the scheme were to span a large number of regions with negative or positive but weakly correlated yields. The alternative of seeking commercial reinsurance seems unlikely given private insurers' reluctance to insure yields against a wide array of perils.

India introduced a national area-yield crop insurance scheme in 1985 (the Comprehensive Crop Insurance Scheme, CCIS), following a pilot phase from 1979 to 1984. The insurance is only available to farmers who borrow credit from financial institutions; and the indemnities are paid directly to the lending institution. Mishra (1994) provides evidence from Gujarat that the insurance did increase lending to small farmers, though repayment rates varied slightly. Unfortunately, the premium rates charged farmers are only a fraction of the rates needed to cover the indemnities paid, and even the small premiums charged (1-2% of coverage) are heavily subsidized by government. As a result, the financial performance of CCIS to date has been disastrous. The CCIS collected Rs 11,961 lakhs ($40 million) of premium during 1985-92, but paid Rs 90,163 lakhs ($303 million) indemnities over the same period (Joshi, nd). Considering that government subsidized about half the premium collected, CCIS paid Rs 15 for each rupee of premium collected from farmers. The insurance program will clearly need to be reformed if it is to become a cost-effective risk management policy.

Another variant of area-based insurance is regional rainfall insurance. In this case the insurance pays out whenever the average rainfall for a region falls below some critical value. Rainfall is easier to measure than average yield, and with modern satellite imagery it is now possible to assess the soil moisture content for a region with the minimum of field inspection. This information could be used to confirm rainfall readings, or even used directly as the insured peril.

Unlike regional yield insurance, rainfall insurance is not tied to the performance of specific crops, Since most farm families rarely depend on a single crop for their total income, the attractiveness of rainfall insurance should depend more on how the total insurance is limited to the occurrence of severe drought or floods, it is likely that in semi-arid or floodprone areas the indemnities would coincide with catastrophic income outcomes for many rural families.

Drought insurance could be marketed rather like lottery tickets, employing lowincome people to sell tickets on a commission basis. Unlike standard lotteries, however, all ticket holders would win a prize in a disaster year, bot no prize would be given in non-disaster years. There is no need to restrict the insurance to farm families, and all types of rural households might find it attractive. This is because a decline in farm incomes in drought or flood years usually leads to a sharp contraction in the rural non-farm economy, and the incomes of many workers and businessmen decline in tandem. Drought insurance faces the same covariability problem as regional yield insurance, but because it is limited to a specific weather peril, it might be much easier to obtain international reinsurance. This prospect could be enhanced if the scheme were run by a commercial bank or insurer within the country.

Schemes of this kind have yet to be tried, and without pilot projects it will be difficult to assess their potential value. Their cost-effectiveness would also need to be compared with alternative means of assisting vulnerable households, such as relief employment schemes, and targeted food rations or income transfers.

3

Water Resources in Agriculture

WATER RESOURCES

The perennial rivers originating in the high Himalayas receive water from snow and glaciers. Snow, ice, and glaciers in the region are approximately equivalent to about 1,400 km^3 of ice. The contribution of snow to the runoff of major rivers in the eastern Himalayas is about 10% (Sharma, 1993) but more than 60% in the western Himalayas (Vohra, 1981).

Because the melting season of snow coincides with the summer monsoon season, any intensification of the monsoon is likely to contribute to flood disasters in Himalayan catchments. Such impacts will be observed more in the western Himalayas compared to the eastern Himalayas because of the higher contribution of snowmelt runoff in the west (Sharma, 1997). An increase in surface runoff during autumn and a decrease in springtime surface runoff are projected in highland regions of south Asia. In addition, the increase in surface temperature will contribute to a rise in the snowline, which, in effect, reduces the capacity of the natural reservoir. This situation will increase the risk of flood in Nepal, Bangladesh, Pakistan, and north India during the wet season (Singh, 1998). No significant changes are projected for annual mean surface runoff in southeast Asia; although an increase during winter and a decrease during summer season is likely.

Reduction of annual surface runoff and increase in temperature will reduce total availability of water both in surface and groundwater in the western region of Asia. The impacts will be greater on irrigation for agriculture and water supply for the industrial and domestic sector. In Saudi Arabia, for example, the total water stress is equal to the total quantities of decrease in groundwater recharge and surface runoff.

As a consequence of global warming, the present distribution of species in high-elevation ecosystems is projected to shift to higher elevations, although the rates of vegetation change are expected to be slow and colonization success would be constrained by increased erosion and overland flows in the highly dissected and steep terrains of the Himalayan mountain range (IPCC-TAR).

There is growing concern that climate change may accelerate the damage to freshwater ecosystems such as lakes, marshes, and rivers. With a rise in temperature, a decrease in the amount of snowfall in the Lake Biwa catchment in Japan is projected-which might exacerbate the process of eutrophication (Fushimi, 2000a). Deterioration of lake water quality also is suggested in Kasumigaura Lake in eastern Japan. The response of lakes and streams to climate change will involve complex interactions between the effects of climate on areal inputs, hydrology, catchments and in-lake processes.

Climate change is expected to affect the boundaries of forest types and areas, primary productivity, species populations and migration, the occurrence of pests and diseases, and forest regeneration. The increase in GHGs also affects species composition and the structure of ecosystems, which, in turn, affects ecosystem function (Schulze, 1994). The interaction between elevated CO2 and climate change plays an important role in the overall response of net primary productivity to climate change at elevated CO^2 (Xiao et al., 1998).

Climate change will have a profound effect on the future distribution, productivity, and health of forests throughout Asia. Warming is expected to be particularly large at high latitudes and so climate change could have substantial impact on boreal forests such as decreasing permafrost areas, improving growing conditions, decreasing areas of disturbed stands and reducing ecosystem quality. Moreover, forest fire is expected to occur more frequently in boreal Asia as a result of increased mean temperature. Pest activity also could increase with a rise in temperature, depending on the age composition of the boreal forests (Alfiorov et al., 1998).

Studies on projected impacts of climate change suggest that northeast China may become deprived of conifer forests. Broad-leaved forests in east China may shift northward by approximately 3° of latitude. These results are based on a 2°C increase in annual mean temperature and a 20% increase in annual precipitation. With an increase in temperature of 2-3°C combined with reduced precipitation as projected for the future in the semi-arid and arid regions of Asia, grassland productivity is expected to decrease by as much as 40-90% (Smith et al., 1996). Approximately 70% of pastures are facing degradation, with dramatic decreases in fodder yield over recent decades in some parts of Mongolia (Khuldorj et al., 1998). Climate change is likely to represent an additional stress to rapid social change in many of Asia's rangelands.

Ecosystems and Biodiversity

There are complex interrelationships and feedbacks between human driving forces and impacts, on the one hand, and climate-and sea level-induced changes and effects on the other (IPCC, 1996). At the interface between ocean and terrestrial resources, coastal ecosystems undergo stress from competing multi-usage demands, while having to retain their functional diversity and resilience in the face of global environmental change (Bower and Turner, 1998). Much infrastructure and economic activity is located in coastal zones, which are vulnerable to sea level rise and more frequent weather events (e.g. cyclones, storm surges). Temperature increases could also adversely affect local flora and fauna of coastal areas, as well as the biological equilibrium of marine life.

Global mean sea level is projected to rise between 9 and 88 cm between 1990 and 2100, with a central value of 48 cm, for the full range of SRES scenarios (Church et al., 2001). The projected central value of 48 cm gives an average rate of sea-level rise of 2.2 to 4.4 times the observed rate over the 20th century. Even with drastic reductions in greenhouse gas emissions, sea level will continue to rise for centuries beyond 2100 because of the long response time of the global ocean system.

A major uncertainty is how the global mean sea-level rise will manifest itself on a regional scale. All models analysed by the IPCC-TAR show a very uneven spatial distribution of sea-level rise. Some regions show a sea-level rise substantially higher than the global average (in many cases more than twice the average) and others a sea-level fall (Church et al., 2001). However, the patterns produced by the different models are not similar in detail. This lack of similarity means that confidence in projections of regional sea-level changes is low, although it is clear that they are important with respect to impacts of sea-level rise on coastal zones.

Climate change will have a wide range of health impacts all across Asia. Although a reduction in health stresses and wintertime deaths is anticipated as a result of less frequent occurrence of extreme cold temperatures in boreal and temperate Asia, an increase in the frequency and duration of severe heat waves and humid conditions during the summer will increase the risk of mortality and morbidity, principally in older age groups and urban poor populations of temperate and tropical Asia (Epstein et al., 1995). In addition, human health is likely to suffer chronically from heat stress related damages. Adverse health impacts result from the build-up of high concentrations of air pollutants such as nitrogen dioxide (NO2), ozone, and air-borne particulates in large urban areas. Combined exposures to higher temperatures and air pollutants appear to be critical risk factors for health during the summer months (Piver et al., 1999).

Global warming will also increase the incidence of some diseases, such as respiratory and cardiovascular diseases, in arid and semi-arid, temperate, and tropical Asia. There is already evidence of widespread damage to human health by urban air quality and enhanced climate variability in Asia. Throughout newly industrialized areas in Asia, such as Chongqing, China and Jakarta, Indonesia, air quality has deteriorated significantly and contributed to widespread heat stress and smog induced cardiovascular and respiratory illnesses in the region (WHO, 2004).

WATER TRANSPORT FLOW FOR IRRIGATION WATER MANAGEMENT

The spatial distribution of crop water requirements can be assessed by combining the information contained in the satellite-derived potential evapotrans piration maps with numerical simulations of soil water balance. Remote sensing data and a digitalized soil map have been used as main input data for the application of a pixel-based, one-dimensional simulation of water flow in the crop-soil system. In recent years there has been quite extensive production of software for the simulation of hydrological processes and a large number of models are currently available. Nevertheless, since consistent efforts are needed for the schematization of each component of the whole hydrological system, surface and groundwater, soils, land uses, hydraulic network, *the* suitability of modelling algorithms to each case study is of particular concern. Several simplifications are introduced in most simulation models applied at regional scale in order to obtain accurate enough outputs without requiring too much input data and computational time. If a satisfactory description of an unsaturated zone is given, a reliable estimate of the soil moisture content in the root zone is output from the simulation run. The information concerning the soil water status allows the actual evapotrans piration to be estimated from its corresponding potential value (derived from remotely sensed data).

Simulating the Farmer's Irrigation Criteria

For simulating correctly the hydrological conditions, irrigation deliveries must be included in water balance calculations. The detection of water stress conditions does not imply necessarily the application of irrigation. At farm level, irrigation could be decided according to completely different criteria, that may not include the soil water status. Often, in the decision-making process, the internal organization of the farm prevails on the actual hydrological conditions. Then the water demand at farm level has to face the structural constraint of the hydraulic network, mostly in terms of available flow rate and pressure head.

In modelling terms, this means that irrigation occurs in a certain plot if a number of conditions are satisfied. These conditions reflect the actual preferences of farmers in the application of irrigation water (time of day, flow rate, required pressure for the delivery system, etc.). In principle, on-demand irrigation systems are regulated by two key factors:

- the depletion of water in the soil root zone;
- the farmer's water application criteria and preferences.

It appears obvious that these two elements mutually influence each other. Nevertheless, the second factor, the farmer's behaviour, seldom appears in simulation studies concerning on-demand irrigation systems, although its relevance to network performance and operation is much greater than water balance considerations. Traditionally, probabilistic criteria have been used for determining flow rates in the design stage of pressurized networks.

These methods, still widely applied in the design stage, do not concern the farmer's actual needs and cannot be used for a correct monitoring of the network performance. One main reason for this neglect could be found in the difficulties in expressing farmers' needs and preferences in an analytical form. However, recent studies have evidenced the possibility of defining farmers' preferences in the management of irrigation water in precise quantitative terms by applying marketing-derived techniques.

In a more simplified manner, the use of questionnaires directed to farmers is able to provide useful indications for defining the rotational intervals and duration. This information can be combined with soil water budget calculations for adequately simulating the operating of on-demand networks. The following aspects should be investigated:

- factors influencing the starting time of irrigation;
- amount of irrigation water applied in relation to crop and soil type;
- flow rate and duration of each irrigation, in relation to farm facilities and available workers;
- possibility and/or willingness to improve irrigation efficiency by reducing volumes applied (e.g. with a maximum amount per season).

From the information obtained from a representative sample of farmers, the relationships between irrigation scheduling parameters and farm characteristics can be defined, also with the support of land use maps resulting from the classification of Landsat images and the digitalized maps of plot boundaries. The farmers' data and the georeferenced archive build up a customized Geographic Information System, which will be indicated in the following with the acronym FIRSA (*Farm IRrigation Scheduling Archive*). Essentially, the FIRSA database contains the following information for each j-th combination of crop type, extension and/or other farm characteristics (machinery, number of workers, etc.):

- the conditions to be satisfied for application of irrigation;
- the average water volumes applied;
- time and duration of applications.

From these data, it is possible to define a function q_j (t), representing the average hourly water flow rate during an irrigation application for the generic plot j^{th}. The FIRSA database is currently under definition for the case study area; for its validation, actual consumptive use of irrigation water was monitored during the 1994 and 1995 irrigation seasons at several nodes of the pipeline network. Since the procedure described here is still under development, it is not possible to provide preliminary results of simulation and comparison with field data. Nevertheless, it is possible to outline some considerations on the actual possibilities of these advanced techniques in irrigation water management.

Although user-friendly interfaces can be built routinely using such a package, its practical implementation requires a large effort and the whole procedure is computationally heavy. The preliminary acquisition of a large mass of data input (soil type, FIRSA database) is rather complex and requires specialized intervention. Furthermore, real-time simulations are hampered by the time required for image processing from the moment of satellite pass. When the preliminary set-up of the whole procedure has been completed, the package can be run by non-specialist irrigation managers for seasonal evaluations concerning the optimal distribution of water volumes, according to the operational features of the hydraulic network and water resources availability. The main attractiveness in using simulation techniques lies in the possibility of tuning one or more variables and evaluating the ensuing results. In this sense, the tool described in this paper may be of great value for the simulation of different scenarios according to alternative water resource management criteria. By so doing, it is possible to identify which strategy maximises the effectiveness of irrigation and the crop production. A map of spatial distribution of such indicators may be derived (D'Urso *et al.*, 1992). Statistics about the performance of irrigation over the entire district enable the identification of critical areas where corrective interventions should be undertaken.

IMPROVE OPERATIONAL FLEXIBILITY AND OVERALL WATER MANAGEMENT

In spite of rigid delivery scheduling, flexibility of on-farm irrigation scheduling can be obtained by providing a storage capacity below the delivery point, so as to compensate for the expected mismatch, especially in time, between deliveries and consumption. Construction costs are not so high, but, in the general case where the reservoir area has to be taken from cultivable land, it should be ascertained that the benefits resulting from more efficient irrigation practices do offset the loss of cropped area. Favourable topography will allow gravity irrigation, otherwise water must be pumped from the reservoir, which entails additional investment and operating costs but may allow the application of water saving irrigation techniques such as drip irrigation. A second way, which may be combined with the one above, is the conjunctive use of groundwater, where available and of a quality good enough for its application as it is or mixed with surface water.

Automation

As it eliminates uncertainties and inadequacies of human interventions, automation of some degree is acknowledged as an efficient means to simplify operation, increase the reliability and flexibility of deliveries and reduce operational water losses. In addition, control methods responsive to demand cannot be implemented without automation.

However, besides unquestionably successful applications, not only in developed countries but also in developing countries (e.g., in north Africa), there are a number of cases where the introduction of automatic control has failed. The reasons for such failures are a design that is too sophisticated with regard to the local conditions and, above all, a lack of sufficient preparation, information and training of the farmers, operators and managers. Automatic control thus passes for costly, cumbersome to understand, master and maintain, and consequently requiring a highly skilled staff with sound knowledge in electro-mechanics, electronics and computers (Horst, 1996).

We will here take issue with this widespread but hasty judgement which prevents developing countries from taking advantage of modernization through well-planned and well-engineered automation. It should yet be clear that, in many cases, efficient automatic control, especially of water levels, can be implemented with very simple structures, such as fixed weirs, or equipment, such as float-operated gates and valves. As for the extra cost, it ranges from around 6 to 8% of the total cost of a new manually-operated system, an amount which can generally be recovered in a few years by water savings, increased yields and improved social order.

Main System Operation through a Proper Design Process

It is beyond the scope of this chapter to detail the control concepts and techniques applicable to canal and pipe systems. We will only say that they cover all possible degrees of operational flexibility and responsiveness to demand. People wishing to go deeper into the matter may refer to papers and publications specially devoted to the subject.

As the above references are mainly related to canal systems, we must point out that pressure pipe systems have the advantages of eliminating seepage losses and being responsive to demand by nature, of course within the limits of their conveyance capacities. However, their application is restrained by a comparatively high cost in the cases of long distances and large flow rates. Nevertheless, a low-pressure piped distribution system, which can be implemented at a reasonable cost, is often an advantageous alternative to tertiary canals.

However, there is no recipe for an ideal control system that would be applicable to all projects. Proper design of irrigation delivery systems is indeed a matter for engineering art and science, subject, however, to a preliminary comprehensive definition of the frame and objectives of the project. As pointed out by Plusquellec *et al.* (1994), satisfactory performance implies that the design should start with the definition of a proper operation master plan that combines and reconciles the various perspectives and expectations of the parties involved, i.e., the farmers, the field operators, the project management, the government and the evaluators. According to the authors, the essential components of such a plan are: level of service anticipated, levels of control in major, minor and watercourse canals and pipes, water ordering procedure, communications, decision support system, data collection and processing, who will do what, plus, at a later stage, specific operating instructions for all control structures.

In a similar perspective, Ankum (1996) presents a classification and a selection process of the control methods applicable to canal delivery systems together with a logic diagram showing the inter-relations between those methods, the modes of water allocation and water distribution, and the decision-making procedures.

Manual Operation and Overall Water Management

The use of an operation simulation model to elaborate setting points for the control equipment of the delivery system does not generally increase the potential reliability and flexibility of the system but may contribute to making them a reality. Numerous user-friendly and versatile simulation programs, that can be run on personal computers by non-specialists in computer science after only a short training, are now readily available (CEMAGREF, 1992; Montañés, 1995). However, due to the hazards of calibration and to the inescapable drift of hydraulic parameters over time, the outputs of the model should be considered with caution and frequently re-validated through feedbacks from the field. On the other hand, if a model can be really helpful for appreciating routing times and adjusting the operating schedules of discharge regulators accordingly, it should not be forgotten that steady manual upstream control can only be based upon level control.

In fact, except for the discharge regulators located at the heads of the main and branch canals, the instructions to operators should simply be to maintain a constant water level upstream of each check structure, something which does not require any simulation. Finally, such models appear to be more suitable for tuning up a system design and testing various operational strategies and procedures than for day-to-day operation. The so-called 'integral' models have a much more comprehensive purpose than just the improvement of delivery system operation. They aim to improve the overall water management in a project, in integrating the various agronomic, economic and operational factors that affect the efficiency of water use. They typically combine simulation of crop water requirements, on-farm scheduling, delivery scheduling and/or main system operation, and can process management indicators.

Three models of this type have been presented:

- The SIMIS program (Scheme Irrigation Management Information System, Morábito *et al.,* 1995) is essentially an information and evaluation system. The reported application to a scheme in Argentina has shown that it can be helpful for the managing staff in detecting discrepancies between plans and reality and in localizing the weak points in the physical infrastructure, operation and management of the delivery system.
- The INCA (Irrigation Network Control and Analysis) software (Makin and Cornish, 1996) is both an information and analysis tool for general management and a decision support tool for day-to-day operation of medium to large-scale irrigation schemes. In particular, it can be used for determining not only optimum delivery schedules according to computed crop water requirements, but also the corresponding operating schedule of the control structures of the main system. Its application to several schemes in Asia, combined with systematic monitoring, has reportedly led to appreciable water savings while at least maintaining and often improving the reliability and equity of deliveries.
- The model package described by D'Urso *et al.* (1996), currently under trial application in Italy, is designed as a decision-support tool combining a mapping of crop water requirements by means of remote

sensing techniques with models simulating the crop-soil system, farmers' irrigation preferences and delivery system operation. Its original feature is that it has been specifically developed for on-demand irrigation systems and therefore includes a module simulating farmers' behaviour.

Such models are undoubtedly powerful and useful tools for project planning, design and evaluation, especially in centrally managed projects and in water scarcity conditions. However, considering their heavy requirements in terms of management organization, extensive data collection and monitoring, calibration, training and installation time, their use for current system operation appears much more questionable.

We are afraid that the applications of such sophisticated models will always remain rather restricted and one may ask whether improving the operational reliability and flexibility of the system through some modernization would not be simpler and more beneficial.

It should be also noted that, in developed countries, where the responsibility for irrigation scheduling is left to farmers, the use of irrigation scheduling models has moved from day-to-day operational applications by the water supplying agencies to on-demand assistance available to farmers from public or private irrigation advisory services (Burt, 1996).

WATER DELIVERY AND IRRIGATION SCHEDULING

As recalled by Burt (1996), reliability of deliveries is a prerequisite for the application of any irrigation scheduling programme. Whatever the delivery schedule, either dictated or resulting from an agreement, either rigid or flexible, it is necessary that water should be supplied to the users in conformity with the expected level of service.

Besides the technical advantage of irrigating in the best possible conditions with respect to possible limitations that are known and accepted by the farmers, reliability of deliveries is an essential condition for establishing, sustaining or restoring a climate of confidence between the water supplier and the water users. Such a climate is indeed a prerequisite for due payment of the water charges, and for the success of any move towards technical or institutional measures aiming at better water management, such as improved delivery scheduling and farmers' participation in operation, maintenance and management through water users' associations.

The requirements for reliability are:

- A design of the conveyance and delivery system that must be not only technically sound from the hydraulic, constructional and operational points of view, but also consistent with both the service required and the local conditions.

- A good-quality construction for both the works and the equipment, all in accordance with the specifications and drawing resulting from the above design.
- Proper operation and maintenance (O&M) of the system, which implies good organization and dedication from the managing level to the field O&M level, as well as enough budgetary resources to cover maintenance costs.

Reliability over time will be all the better since provisions have been made at the design and implementation stages to accommodate any changes in the physical condition of the system or the service required.

Equity Versus Flexibility

Equity, i.e., deliveries in proportion to the cropped areas served, has long been, and still is, the leading principle of delivery scheduling in most projects, especially run-of-the-river schemes and where water is scarce. This principle is, moreover, in accordance with the traditions or laws governing water rights in many countries. Such systems can be implemented and operated with very simple non-adjustable structures and equipment and no communication facilities. Operational and managing requirements are thus reduced to a minimum. In fact, theoretical equity is seldom a reality. Two categories of reasons can be considered:

- It is virtually impossible to design non-adjustable outlets that work properly over a wide range of flow rates and levels. In addition, the functioning of the dividing structures is inevitably affected by the drift of canal hydraulic parameters over time (Plusquellec *et al.*, 1994).
- Especially in the case of loose management, the actual sharing out of water may deviate widely from the official scheduling, as a result of the dominant position of some farmers, non-registered transactions on water rights and land ownership, or uncontrolled extension of irrigated area (Bandaragoda, 1996).

On the other hand, except for mono-cropping and cases where water requirements of different crops are similar, equity-based delivery scheduling is obviously inconsistent with crop-based and water-saving irrigation practices.

However, case studies presented by Bandaragoda (1996) show that attempts to move from equitable, or so presumed, delivery schedules to more flexible crop-based schedules may fail because of resistance to change from not only the managing agencies but also the users. Whether the decision on when and with how much water to irrigate is up to the supplier or to the users, flexibility of deliveries is still acknowledged as a precondition for applying efficient methods of irrigation scheduling.

Adjusting water supply to irrigation water needs varying over time and space obviously implies that deliveries to each farmer or group of farmers could be varied in frequency, flow rate and/or duration and, consequently that, the whole delivery system could be operated in a flexible and efficient way. This can be achieved by applying relevant design concepts, equipment, and operational tools, referred to in the following section.

However, as developed by Burt (1996), delivering water with a high degree of flexibility and reliability depends not only on technical means, but also requires:

- a decentralization of decisions and responsibilities based on the breakdown of the delivery system into 'levels of service' (e.g. main canal level, secondary canals level, etc.), each one being responsible for providing a well-defined service to the next lower one;
- the institution of seasonal or yearly water allocations, so as to avoid or, at least, minimize the water wastage and inequities that may result from flexible schedules, and especially from on-demand schedules (cf. Horst, 1996).

Finally, applying flexible deliveries requires a number of technical, social, educational, organizational and financial conditions that are seldom met in developing countries, as set out by Horst (1996) and Mangano (1996). Both have reservations about the suitability and applicability of flexible delivery schedules in such countries. They consequently recommend sticking to the simplest scheduling and operating modes, and ensuring that they are actually applied and trying to obtain the most from their low flexibility. We can only agree upon with realistic approach, based on the fact that 'successful irrigation scheduling must be simple to implement and easy to understand from both the farmer and project management personnel standpoints' (Burt, 1996) and that the excess deliveries resulting from simplified schedules are likely to be compensated for by the decrease in operational losses resulting from simpler operation (Horst, 1996).

However, considering the absolute necessity for water savings and sustainable food production, we are convinced that improving the flexibility of delivery schedules and of delivery system operation to some degree should be a permanent concern, at least for the longer term.

AGRICULTURAL DIVERSIFICATION

India's future agricultural strategy must also be oriented to the need for agricultural diversification. India's foodgrain production capacity has increased significantly over the years and there is evidence that household consumption patterns are changing away from foodgrain towards higher value crops such as vegetables, fruits, milk, eggs, etc. Future growth in agriculture must come from diversification into these non-food grain areas and this will pose a special challenge because marketing these perishable products is much more complicated than marketing food grains. Horticulture development is currently constrained by poor marketing arrangements. The gap between prices received by the farmers and those paid by urban consumers is large, reflecting inefficient marketing arrangements. Horticultural produce is typically collected from farmers by market agents, who sell it organised markets established under the Agricultural Produce Marketing Acts. Unfortunately these markets are controlled by a few traders and operate on highly non transparent basis. Facilities for grading and handling are poor and methods of price discovery in the markets are nontransparent. Wastage is high owing to poor logistics and the absence of cold chains. The net result is much lower realisation by the farmer.

It is necessary to amend outdated laws restricting the establishment of markets to allow cooperatives and private entrepreneurs to set up modern markets with grading facilities, cold storages and transparent auction procedures. Half a dozen states have already amended their existing laws on agricultural marketing to allow such markets to be established and a dozen others are in the process of doing so. These changes are being resisted by those who control the existing structure but this opposition will weaken over time. Contract farming is another innovation that has been introduced in many states and could accelerate diversification. India's laws on agricultural land do not allow corporate bodies to purchase land and operate large-scale farms – a national policy to prevent displacement of a large number of small farmers. – but corporate buyers who know what is needed in export markets, in high end domestic markets or in agro-processing, can engage in contract farming to procure high quality produce. Buyers select areas suitable for the crops they are interested in and organize farmers to produce these crops under contract, while providing planting material of the right quality and technical supervisory. The process enables the farmer to eliminate marketing risk while allowing the corporate buyer to ensure quality supplies by selecting planting material, and provide access to scientific advice on disease and other types of stress.

The development of agro-processing will be a spur agricultural diversification and the government is paying special attention to this area. At present, the proportion of India's agricultural output that is processed is very small compared with that in most developing countries and the demand for processed food is bound to increase as incomes rise. There are several obstacles to the more rapid development of food processing. Taxation structures often discriminate against food processing because processed food is the first stage for application of indirect taxes and the absence of tax rebate on taxes paid on inputs means the effective tax on value added is very high. Another impediment is the reservation of certain categories of products for small scale production. The absence of a modern food processing law has meant that this sector is governed by multiple laws, making it difficult to operate effectively. An Integrated Food Processing Law has been introduced in Parliament to replace and the passage expected in the current year will make a qualitative difference to the operating environment.

APPLICATION OF SPM TO WATER RESOURCES

Throughout the implementation of the project on "Capacity-building in Strategic Planning and Management of Natural Resources in Asia and the Pacific", it was recognized that water resources development planning is a continuing process and the introduction of SPM into the national development process would thus need to be integrated into the ongoing efforts to improve water resources development planning. In that context, the principles for application of SPM to water resources were drawn up on the basis of the importance or necessity of SPM, key elements or components of SPM, and initiation and implementation of SPM.

SPM for Water Resources Development

Throughout the implementation of the project on "Capacity-building in Strategic Planning and Management of Natural Resources in Asia and the Pacific", all participating experts and institutions recognized the importance and necessity of the adoption of SPM approaches in water resources management to ensure effective integration of water resources management into the national development process, to achieve efficient allocation of water resources against competing uses and to achieve consensus building and sustainable results of development.

It was further recognized the importance of SPM in promoting good governance in water resources management and in capacity-building and improvement of management to cope with the increasing scarcity of water resources in the subregion. The participating experts also identified different priorities for the application of SPM to water resources at the national, sectoral and organizational levels.

Development of IWRM Plans

SPM approaches were recognized as instrumental to the application of the concept of IWRM to support the socio-economic development process.

While IWRM is acknowledged as a process, SPM was recognized as a powerful tool to enable priority activities of water resources management to be effectively integrated into the socio-economic development process through the following aspects:

- SPM enable priority actions to be strategically implemented.
- SPM is a process based on the scenarios analyses for the future, allowing uncertainties of the future to be incorporated into the planning process.
- SPM helps to regularly assess persisting issues.
- SPM promotes accountability, especially the core groups mainly responsible for the implementation of SPM.

It was considered that application of SPM to IWRM is the only choice to meet the urgent needs of socio-economic development when countries face serious water issues. It was also recognized that the application of SPM to development of IWRM plans would facilitate the mobilization of political will and commitment of key stakeholders, promote resolution of conflicts and ownership, and enable development of practicable plans of action.

The participating experts also drew up the following experiences of their application:

- SPM can help train new generations of water professional who would understand the challenges multidisciplinary approach needed for IWRM.
- SPM helps dissect major issues into manageable issues/matters.
- SPM develops capacity to understand the pressing issues in terms of socio-economic aspects and other frameworks.
- SPM helps establish a core group comprising members from different agencies.
- SPM promotes pilot river basins.
- SPM is the effective tools to attract the assistance from donors and shows clear areas of responsibility, where stakeholders are participating.
- SPM can provide catalytic elements for change management and transform transitional institutions to think in terms of real time situation.
- Political will/government support and financial resources are needed for a successful SPM.
- SPM helps in raising awareness among various stakeholders.

Expectations from SPM

The participating experts in the Concluding Workshop mentioned above also identified the following expectations from the application of SPM approaches to water resources management as a possible checklist:

Mechanism of Work

- Use SPM to develop network: SPM can be used as a tool generate consensus building, such as possible guidelines to develop action plans for food security and poverty alleviation
- Strategic Plan of Management of Water Resources: must take into cognizance Water Resources assessment within a river basin. In many countries, economic development is based on administrative land. What are areas of collaborative planning that can be developed, particularly economic, among stakeholders
- Increase networking through partners and among stakeholders
- Better decision-making in the process of water resources management
- Task force for collaboration is also essential
- Integration of IWRM to national social and economic development strategy and action plan
- To set up new legal institutions, needed for better management
- Building up of an effective coordinated block of all water related organization
- Improved water resources management structure of the Government

Change in Mindsets

Successful application of SPM approaches to water resources management is expected to result in the change of mindsets in the following areas:

- Development concept: socio-economic
- Water concept: IWRM, water utilization, ecological rehabilitation, flood mitigation
- Implementation: mobilization of support and commitment of policy decision-makers
- Foundation of water management: improvement in the understanding of the "reality" of water management systems as the basis for integration to form the basis to see how things actually work, i.e. going beyond institutional and other frameworks
- Efficiency of water resources management: change in mindsets to conserve water resources

Mobilization of commitment for support and participation

- Instill in every stakeholder's mind/citizen on future scenario based on vision developed by them

- Enhance organizational effectiveness of the water resources coordination system and promote capacity-building
- Set up and develop the water market
- Civil society involvement in water resources management.

Implementable Programme of Action

- Ownership and sense of belongings of plan of actions developed by each basin
- Coordinated manner between social, economic and environmental aspects
- SPM should help us in identifying priority actions also at sectoral level
- The devil is in the details; having good SPM should help us go into the details of what to do
- Collaborative planning to achieve IQRM focusing on elemination of side effects due to sectoral focus
- To use the limited resources (water, human, money, etc.) in the best way
- Identification of areas need to be developed/improved

Vision Statement of the Ministry of Water Resources

In order to carry out the new development perspective of the Central Committee of the Communist Party of China, the Ministry of Water Resources has changed its thinking on implementing sustainable approaches to water resources management. The essence of the new viewpoint is harmony between humans and nature, requiring changes in traditional ideas and behaviour. The first change is to shift from the concept that water cannot be exhausted to the concept that freshwater is a limited and vulnerable resource; from the view that water is a free natural resource to that of water being an important strategic economic resource.

The second change is a shift from unbounded development and use of water resources, without consideration being given to ecological needs, to paying careful attention to ecological use and rationally arranged domestic, productive and ecological water use. The third change is from simply emphasizing flood control and prevention to a coordinated relationship between humans and nature in order to preserve the necessary space for flood flows and storage, and to make use of floodwater as much as possible with the prerequisite of ensuring safety against flooding.

The fourth change is a shift from biased emphasis on development, utilization and harnessing of water to coordinated development, utilization, harnessing, allocation, conservation and protection, with the focus gradually being moved to rational allocation, high-efficiency utilization and effective protection. The fifth change is a shift from the separation between water resources and the economy and society, to basing population growth, economic structure and layout of productivity, development scale and speed on the carrying capacity of water resources, while seeking the coordinated development of population, resources, economy, society and environment.

The sixth change is a shift from only depending on engineering measures to integrating engineering and legal, administrative, economic and technical measures, and from engineering planning to strategic integrated water resources planning. The seventh change is a shift from only using surface water or groundwater resources to jointly using surface water, groundwater, rainwater, floodwater, low-quality water, reclaimed wastewater, seawater, desalinized seawater etc.

The eighth change is a shift from water supply management that only pays attention to satisfying demand to management that applies multiple measures to rationally check demand. Establishing a water-saving society is a new task of the Ministry of Water Resources. The ministry will widely publicise water saving in society, encourage the agricultural and industrial sectors to adopt new methods of saving water, and promote the use of new equipment for domestic water saving.

Combined with environmental considerations, the Ministry of Water Resources will strengthen water resource protection in its efforts to control water pollution. The ministry will also promote efficient water resources use further through water allocation. It is currently compiling comprehensive water resources planning, which will fully consider water usage in the various watersheds and by water use departments and the domestic sector as well as in production and the ecology.

In order to promote water-use efficiency; the Ministry of Water Resources will take into full consider the use of market mechanisms in the future. At present, water rights and water markets are not in good form; therefore, the ministry will develop them and introduce them into water resources allocation.

Mission Statement of the Ministry of Water Resources

The Ministry of Water Resources is the department of the State Council that is responsible for water administration. In accordance with the stipulations of the State Council, the Ministry of Water Resources has been given the following mandates.

(a) Formulation of water-related policies, development strategies and medium-term and long-term development plans, including water conservation and demand management policies;

(b) Drafting and implementation of enabling legislation and the design of water-related regulatory frameworks. This includes the implementation of the water-drawing permit system and the water resource fee system, and the mediation and arbitration of intersectoral and interprovincial water disputes;

(c) Implementation of integrated water resources management, including atmospheric water, surface water and groundwater. This includes:

   (i) The formulation of national and interprovincial development plans for water supply and demand, and schemes for water allocation

   (ii) Supervision of the execution of the above plans and schemes

   (iii) The assessment of water resources, flood risk and flood mitigation measures in relation to the overall planning of the national economy, urban planning and major construction projects

   (iv) The publication to national research and development in hydrology

   (v) Guidance of national research and development in hydrology

(d) The formulation of water resource protection plans in accordance with related national laws, regulations and standards concerning resource and environment protection; the demarcation of functional water areas and the control of wastewater discharges into potable water areas and other water areas; monitoring the quantity and quality of water in rivers, lakes and reservoirs; and review and approve the pollution loading capacities of water bodies with proposed limits on total wastewater discharge;

(e) The formulation of economic regulatory measures for the water sector; the enforcement of macroeconomic regulations on the utilization of funds within the water industry; the provision of guidance on economic activities related to water supply, hydropower and diversified development within the water sector; and the provision of recommendations on economic regulation of water pricing, taxation, credit and financial affairs;

(f) Drafting and reviewing of proposals and feasibility study reports on large and medium-sized capital construction projects in the water sector;

(g) Drafting and supervision of the execution of technical standards for the water sector as well as specifications and codes for water works; the implementation of key hydro-science research projects, and the popularization and dissemination of water-related technologies;

(h) The organization and direction of management protection for hydraulic facilities, water areas, dykes and coast lines, and the regulation, reclamation and development of major rivers and lakes as well as beaches; handling foreign affairs in relation to international rivers between China and its neighbouring countries; the organization of construction and management of key controlling and interprovincial hydro-projects; and the organization and direction of monitoring and management of the safety of reservoirs and dams of hydropower stations;

(i) Provision of guidance on activities related to rural water resources; the organization and coordination of capital construction of farmland drainage and irrigation, rural electrification, and water supply for townships and villages;

(j) The organization of water and soil conservation nationwide, including the formulation and development of engineering measures for water and soil conservation; and organization of monitoring and overall prevention and control of soil and water losses;

(k) Responsibility for activities concerning science, technology and foreign affairs that are related to water resources, including the provision of guidance in the development of a competent workforce for the water sector;

(l) Responsibility for the day-to-day work of the State Flood Control and Drought Relief Headquarters. This includes the organization, coordination, supervision and direction of nationwide flood control as well as the execution of flood control and drought prevention operations in major river basins and key water projects; and

(m) Other duties and responsibilities assigned by the State Council.

Water Resources Coordination Committee

The abundant water sources in the Lao People's Democratic Republic have the potential to support socio-economic development, especially the hydropower and irrigation subsectors. The hydropower potential of the country is large compared to other countries in the lower Mekong River basin, providing an opportunity to earn foreign revenue. The hydropower sector also has the ability to develop rapidly; it increased its production about five times from 247 million kWh in 1976 to 1,187 million kWh in 1999 when it exported 473 million kWh. The Government has given high priority to investment in the irrigation subsector as agriculture forms the foundation of national economic development, and is necessary for food stabilization. In addition, some 85 per cent of the population lives in rural areas. From 1976 to 2000, the area of irrigated dry season rice increased about 40 times from 2,700 ha to 110,000 ha. The irrigation subsector also significantly increased the average yield of rainfed paddy rice from 1.43 mt/ha in 1976 to 3.27 mt/ha in 2000. However, water resources development is still at a low level: the irrigated area is only 20 per cent of the national paddy area while hydropower production is still at only 2 per cent of its potential of 30,000 megawatts.

Development in other sectors is still at a low level compared with hydropower and irrigation. In addition, several problems are faced by the water sector. These include:

(a) Unusual rainfall patterns in some years, plus high evaporation, flooding and drought in some of the main agricultural areas of the country;
(b) The impact of shifting cultivation on water resources, although this activity has been significantly reduced; and
(c) Conflicts of interest for management within the sector since most water subsectors are still responsible for multiple roles of regulator, manager and service provider.

As the Lao People's Democratic Republic is a least-developed country with rich water resources, the most important challenges for WRCC in carrying out its coordinating role include:

(a) Strengthening the legal framework for effective and harmonious integration of water resources management, development and protection activities into the socio-economic development process of the country, in particular to meet national priorities;
(b) Enhancing and consolidating the existing systems and foundation to operate, maintain and rehabilitate facilities safely, reliably and efficiently in order to protect the investment for public benefit; and
(c) Prioritizing the capacity-building needs in order to enhance the organizational capacity and effectiveness of the water resources coordination system.

Water Resources Utilization and Challenges

Myanmar is an agricultural country with an abundance of water resources. The agricultural sector is the most basic economic of the State as well as the main source of livelihood in rural areas, since the rural population represents some 70 per cent of the nation's population. The development programmes for the agricultural, livestock breeding and fisheries sectors are included. At present, the State has been systematically disseminating advanced techniques and support for the development of the nation's economy. Irrigation facilities have been increased since 1988 to reach 163 in August 2004. Dams are now irrigating more than 1 million ha of farmland. In addition to the dams, river water pumping stations, underground water tapping stations and small dams have been built throughout the nation. A total of 265 river pumping projects are irrigating about 150,000 ha of cultivated land. In addition, 7,478 tube wells have been provided to irrigate 36,000 ha of farmland.

Tributaries originating in the western hill region and southern part of the country constitute some 10 per cent of the catchment areas and surface runoff. The hydropower potential of these tributaries is considerable. According to studies by the United Nations and other sources, the hydropower potential of Myanmar is estimated to be as much as 40,000 MW. By 2002, 35 hydropower stations (including 15 medium-scale projects) had been completed with a total estimated generated power of 390 MW, which is almost 1 per cent of potential generated power in Myanmar. The development of the electrical power sector thus can provide a significant contribution to the socio-economic growth of the country. Therefore, electric power projects are being implemented wherever possible to fulfill the electricity needs of Myanmar.

The Development Affairs Committee under the supervision of the Department of Development Affairs has implemented 15,715 projects to supply drinking water to 10,602 villages, using funds form the Government, the Department of Development Affairs and donors at home and abroad. The projects have been implemented with participation by international and non-governmental organizations. The water sector faces several problems including unusual rainfall patterns in some years, flooding and drought in some of the main agricultural areas of country, the impact of shifting cultivation, illegal logging in water resources areas as well as management conflicts of interest and a lack of coordination within the agencies. The most important challenges include:

(a) Strengthening the legal framework to ensure effective and harmonious integration of water resources management, development and protection activities into the socio-economic development process of the country;
(b) Enhancement and consolidation of the existing systems;
(c) The operation, maintenance and rehabilitation of facilities safely, reliably and efficiently; and
(d) Prioritizing capacity-building needs in order to enhance organizational capacity and effectiveness of the water resources coordination system.

Socio-economic Development Context

Agriculture (including crop production, livestock, fisheries and forestry) is the most important sector in the country's economy because:

(a) It is the main source of livelihood for about 70 per cent of the population, who live in the rural areas;
(b) It accounts for about 64 per cent of the labour force;
(c) It contributes about 41 per cent of export earnings; and
(d) It contributes about 42 per cent of the gross domestic product (GDP).

Due to the importance of the agricultural sector, the Government has accorded high priority to its development, and numerous irrigation facilities have been implemented during the present decade for irrigation and water supply to monsoon and summer paddy crops. Total irrigated area increased from 1.1 million ha in 1980-1981 to some 2 million ha in 2000-2001. Apart from the agricultural sector, the hydropower subsector is also the most important in terms of economic development and investment. Dry-season irrigation, especially from river pumping projects, has been successfully increased; as a result, power demand is also increasing annually.

The Government is making every effort to be able to catch up with development in other countries around the world. The power generation sector plays a very important role in those efforts and it will contribute considerably to socio-economic growth. Therefore, electric power generation projects are being implemented wherever possible in order to meet the demand for electricity supplies. The Government is implementing plans for the national grid and additional power supply projects. There are 11 major projects being implemented for the power grid and a further 19 projects are planned. Three regional projects are being implemented, with an additional 10 planned for implementation. Myanmar is one of the richest water resources countries in the Association of South East Asian Nations. In view of its current level of water utilization, Myanmar can therefore sell electricity and drinking water if there is a consensus for Global Water Partnership on a regional basis (GWP-SEA).

Threats in Water Resources Development

Myanmar is rich in water resources. The total utilization of the nation's water resources is only about 5 per cent or 45 million acre-feet (56 km^3). It is clear that the physical potential for further development of water resources in Myanmar is substantial. Myanmar has now reached a major turning point in the use of water resources for all-round development of the country. Providing innovative and integrated solutions for sustainable management of water resources to meet national development needs has become a necessity. At the same time, the increasing water use will need careful management to prevent unwanted effects.

Sedimentation is one of the major adverse effects of storage dams and in the lower courses of rivers. Mining and deforestation along the upper reaches of river basins cause serious watershed erosion problems. Transported sediment is reducing the storage capacity of reservoirs and the bed level in the lower reaches of rivers is rising. Consequently, flooding occurs and navigation faces serious problems. Although some nutrients and some sediment are needed to support the aquatic environment, the Government is emphasizing the implementation of the terrace farming system to reduce shifting cultivation.

The development of industry and increasing population density will cause increasing river pollution and health risks for people living close to the rivers. Careful management of groundwater extraction is also required in order to avoid pollution.

Flooding is a major issue, with damage occurring both in urban and rural areas. On the other hand, flooding brings some benefits as it creates rich floodplains and replenishes soil moisture. Flood plain management could be considered as a way to reduce flood damage, with sensitive development being encouraged away from flood-prone areas. The State has given higher priority to providing adequate support and assistance for further development of the fisheries industry in order to enhance food security while protecting the environment. Changes in land use, declining water quality, the construction of dams and other barriers to migration and high levels of capture are likely to be responsible for declining of fish populations. Therefore, increases in water bodies for aquaculture systems such as reservoirs, river courses, fishponds etc. should be temporarily stopped.

WATER MANAGEMENT IN RAINFED AREAS

About 60 percent of India's cultivable area will remain dependent on dry land farming even after all irrigation potential is fully exploited. Productivity growth in these areas is obviously critical for rural income growth and poverty alleviation, and it depends critically upon better moisture conservation and the development of varieties suited to dealing with moisture stress.

Schemes for water retention, moisture conservation and groundwater recharge have been implemented for many years in India but with mixed results. Experience suggests some pointers for the future. Greater use of technology inputs can help a great deal. Satellite mapping by the Indian Space Research Organisation (ISRO) has been particularly helpful in planning watershed management schemes to achieve optimal results. It is also important to adopt a holistic approach. For example, if deforestation problems upstream are not tackled, water retention structures downstream will quickly silt up. Community participation is critical to impart ownership and ensure an acceptable distributional outcome. In the past, these multiple factors were not effectively integrated into watershed development schemes. Now a National Rainfed Area Authority has been proposed to help coordinate the work of different implementing agencies.

The cost of treating rainfed areas to ensure optimum use of available water is approximately Rs.10,000 per hectare and the untreated area is about 80 million hectares, yielding a total cost of approximately US$20 billion. If this amount is added to the cost of irrigation development and the target to be achieved over a 10-year period it would require a doubling of public investment in irrigation.

Other Inputs

Increasing agricultural productivity also depends on the efficient delivery of several other inputs. The quality of seeds and planting material needs to be greatly improved, and this calls for strengthening the research effort to make it more effective. Two expert committees have recently reported on how to restructure the agricultural research system to make it more results oriented and their recommendations are under consideration. The system for producing and marketing certified seeds of existing varieties at reasonable prices also needs to be improved. Seed replacement rates in most parts of the country are only one-third to one-half of what they should be, a situation which reflects partly a lack of knowledge of the importance of seed replacement and partly also a lack of availability of reliable seeds.

There is evidence that the use of fertilizers is at present highly imbalanced suggesting that scientific application of fertilizers holds potential for raising productivity. Nitrogen fertilizers are oversubsidized compared with phosphorus and potassium fertilizer. The structure of fertilizer subsidies should be rationalized to avoid excessive and wasteful use of nitrogen fertilizers. Lack of knowledge of micronutrient deficiency in the soil is also a serious problem. There is need for much more extensive soil testing to encourage balanced application of nutrients needs. Underlying these problems is the deterioration of the extension services, which makes it difficult to disseminate best farming practices. Strengthening the extension system therefore needs special attention.

4

Rainwater Harvesting

Rainwater harvesting is the accumulating and storing, of rainwater. It has been used to provide drinking water, water for livestock, water for irrigation or to refill aquifers in a process called ground-water recharge. Rainwater collected from the roofs of houses, tents and local institutions, can make an important contribution to the availability of drinking water.

Water collected from the ground, sometimes from areas which are especially prepared for this purpose, is called Stormwater harvesting. In some cases, rainwater may be the only available, or economical, water source. Rainwater harvesting systems can be simple to construct from inexpensive local materials, and are potentially successful in most habitable locations. Roof rainwater can be of good quality and may not require treatment before consumption. Although some rooftop materials may produce rainwater that is harmful to human health, it can be useful in flushing toilets, washing clothes, watering the garden and washing cars; these uses alone halve the amount of water used by a typical home. Household rainfall catchment systems are appropriate in areas with an average rainfall greater than 200 mm (7.9 in) per year, and no other accessible water sources (Skinner and Cotton, 1992).

There are a number of types of systems to harvest rainwater ranging from very simple to the complex industrial systems. The rate at which water can be collected from either system is dependent on the plan area of the system, its efficiency, and the intensity of rainfall (i.e. annual precipitation (mm per annum) x square meter of catchment area = litres per annum yield)... a 200 square meter roof catchment catching 1,000mm PA yields 200 kLPA.

HISTORY OF HARVESTING

History of Harvesting India has had a tradition of water harvesting which is more than two millennia old. It is in the nature of Indian ecology that forced Indians to develop the art of water harvesting. Though Indian gets a high amount of rainfall, it is not evenly spread across the year. Most of the time, even in a normal rainfall year, the country faces a drought. It is only on a few days and a few hours that all the rain comes down 'pouring cats and dogs', as the British discovered when they came to India. Eminent meteorologist, P.R. Pisharoty says that in most parts of the country, there is precipitation during not more than 50 days. On the days when rainfall does occur, it does not fall over a period of 24 hours. In fact, heavy showers are common. Most of the country receives rain for just about 100 hours each year. The remaining 8,660 hours in a year get no rain.

Not surprisingly, any water that falls from the heavens or flows past in a stream has been harvested in Indian since antiquity. Says archaeologist B.M. Pande, "Evidence of this tradition can be found in ancient texts, inscriptions, local traditions and archaeological remains. There is some evidence of advanced water harvesting systems even from the pre-historic times. Famous Hindu texts like the Puranas, Mahabharata, Ramayana and various Vedic, Buddhist and Jain texts contain several references to canals, tanks, embankments and wells." The first reference to the hydrological cycle in the world comes from the Chandogya, one of the principal Upanishads (the philosophical reflections of the Vedas) which points out: "The rivers all discharge their waters into the sea. They lead from sea to sea, the clouds raise them to the sky as vapour and release them in the form of rain..." This reference shows that as early as 1000 B.C, attempts were being made to interpret and explain recurrent natural phenomena on the basis of direct experience. Of particular historical significance is the evidence that has recently come to light in Dholavira major site of the Indus Valley Civilisation, dating back to the third millennium BC. The site is located in the Great Rann of Kutch, and one among the five largest Harappan cities found so far. Dholavira discovered in the 1960s, lies in an arid area which gets an average of 260 mm. annual rainfall. Their are no perennial sources of water in the form of lakes or rivers. Subterranean water is, by and large brackish and saline. The inhabitants of Dholavira, therefore, created several reservoirs to collect the monsoon runoff flowing down the flanking streams of the Manhar and Mansar. Stone bunds were raised across them at suitable points to divert the flow of water into a serried of reservoirs which were dug out in the sloping areas between the inner and outer walls of the Harappan city. Likewise, a network of drains crisscrossing the citadel was also laid out to collect rainwater.

In his book Arthasastra, written in the fourth century BC, Kautilya-the mentor of India's first emperor, Chandragupta Maurya (321-297 BC), who was responsible for throwing the Greek invades out of India-in the post-Alexander period, speaks at length about the need to construct irrigation systems and reservoirs and sets out rules for collecting revenues and for fines to be imposed on those who do not observe state rules set for the good management of the water systems.

This tradition was carried through over centuries. The famous Vijaynagar kings of medieval south India (1336-1564 AD) placed great importance on developing irrigation facilities for agriculture improvement. Emperor Krishna Deva Raya (1509-1530), the greatest of the Vijaynagar rulers, would pointed out that the extent of a state is the root cause of its prosperity, and if is small, its prosperity will increase only when tanks and irrigation canals were constructed and favour was shown to poor cultivators in matters of taxation and services. The Vijayanagar kings, therefore, constructed irrigation tanks and canals themselves, though more rarely. But, more importantly, they encouraged private initiative in irrigation development in different ways.

The early British observers were full of praise for the irrigation works they saw. About the irrigation policy of the Vijaynagar kings, C.S. Crole, the author of The Manual of Chengleput District, 1879 said: "Almost every catchment basin, however, small, still bears traces of having been bunded across in many instances this was done in order to secure a crop of paddy on a few acres of ungenerous soil, to which all the fostering care of the British administration has failed to cultivation to return. Large and more expensive projects were not neglected. Even some of witness to the enlightenment of those Hindu kings, while the absence of scientific instrument those remote times compels the astonishment of the beholder."

Colonel Thomas Munro (later knighted and made the Governor of Madras in May 1820), noting irrigation system under the Vijaynagar kings in the area around Vijayanagar, said: "To attempt construction of new tanks is perhaps a more hopeless experiment than the repair of those which been filled up (through siltation), for there is scarcely any place where a tank can be made advantage that has not been applied to this purpose by the inhabitants.

A Trove of Technology

Over the centuries the Indian people developed a variety of water technologies based on the ecological conditions and needs. India has an extraordinary diversity of agro ecological ranging from the hot desert of Rajasthan to the cold desert of trans-Himalayan Ladakh, from the temperate Himalayan mountains to the high tropical mountains in the south, and interspersed various hill and mountain ranges, plateaus and the unique Indo-Gangetic Plains which are flood-prone than any other part of the world. Traditionally, wherever there were streams, especially in the hill and mountain regions of India, people developed techniques to divert the water, with help of simple engineering structures, into artificial channels that would take the water to agricultural fields. When streams became bigger and turned into rivers, the engineering also be' more sophisticated and diversion systems became bigger and larger.

In the arid and semi-arid regions, where the streams were more seasonal, the diversion were first directed into a storage structure-variously called a zing in Ladakh, an ahar in south East or a kere in Karnataka-so that water could be used in the dry period for human and consumption, and for agriculture. Stream flow in the desert of Ladakhi, as it is dependent on melt, is almost non-existent in the morning but heavy in the evening, when cultivators cannot. Therefore, the Ladakhis traditionally divert the evening flow into a small reservoir to use it morning. In certain villages of the Eastern Himalayan states of Nagaland and Arunachal Pradesh villagers take the diverted channel through a cattleshed so that the water can pick up rich nutrition before it reaches the fields. The Angami Nagas ensure that the forests above the main stream never cut so that the stream water remains full of nutrient-rich forest humus. Not all storage structures are riverbed or streamfed. Many of them simply collect water running catchments area to be stored for later use. But in the southern state of Tamil Nadu, a big stream often diverted to feed a chain of 25-30 tanks in sequence. Because this chain of tanks, also c 'system tanks', is served by a stream collecting water over a large catchments, it is tradition considered much more desirable than a tank with a single, small catchment. Because of the variability in rainfall in both space and time, agricultural lands served by system tanks were more expensive and prized than lands served by stand-alone tanks. The Palar anicut (dam) for example, supplies water to 317 tanks, irrigating an irrigation command of about 32,000 I" North Arcot and Chengalpattu districts.

Some tanks are supplied directly by the channels co from the Palar river but several others depend mainly on the surplus flows of upstream tanks profusion of tanks in Ramnathapuram district of Tamil Nadu can even be seen from a space satel. In several parts of India, people would build a dam across a seasonal channel to capture the n But these structures unlike normal dams were used to moisten the soil so that the post-mor crop would be assured on the rich, moist soil of the tank-bed itself. In the haveli system of Madhya Pradesh, the soils and traditional crops were such that the farmers found it useful to store rain\ in the agricultural fields itself.

The fields would be embanked and farmers would arrangement amongst themselves to allow rainwater runoff to flow from one field to another. " the collected water would seep down into the soil and give it enough moisture to take a good the following dry period. In the dry areas of Rajasthan, people traditionally practiced conjunctive use of surface water ground-water.

They invariably built structures like wells and stepwells-elaborately carved wells v\ flight of stairs leading down to the water-below tanks and other types of water storage structure this way, during the dry season when the tank water dried up, people could at least harvest ground-water to meet their drinking water needs. In other places, people would make wells ii tank bed from which they would collect ground-water for drinking.

In many places, the surface collected in the tank would not be used for drinking if clean ground-water was available from well; stepwells. The most outstanding example of this conjunctive use of water is in the city of Jodhpur. Or fabulously rich desert fort, the Jodhpur fort is situated at the edge of a rocky plateau. The kings had built a series of canals to collect the runoff from the plateau and channel it into se tanks built in and around the fort.

For the houses that came up on the incline down the numerous stepwells and wells were constructed to capture the seepage from the reservoirs its 500-year old history, Jodhpur has witnessed many droughts but never a desperation for water. In the flood plains too, people developed ingenious techniques to use the menacing floodwaters just to irrigate their fields but also to fertilise them and control diseases like malaria by making the fish in the floodwaters to eat away the mosquito larvae. India's richest agricultural area in British India, the flood plain of Bengal, which sits astride the most floodprone region in the world developed an extraordinary mechanism for harvesting the rich floodwaters in a measured constructive way.

With agricultural production declining rapidly in the once prosperous Bengal-for instance, the p yield in Burdwan district dropped from an average of 1.12 tonnes per acre in the period 1871-0.75 tonnes in 1924 and with the net cropped area dropping from 68 per cent of the district's land area in 1871 to 38 per cent in 1921-the British government invited Sir William Willcoc British irrigation expert who had earlier worked in Egypt and Iraq, to advise it on development in Bengal. In a series of lectures delivered in Calcutta in the 1920s, Willcocks, which known for first talking to the local natives before making up his mind, stunned everyone by an that the absolute best that the government could do was to revive the ancient flood irrigation of the region.

When Willcocks tried to plan a system of irrigation canals for the Bengal countryside, he astonished to find that every 'dead river' on the map was an appropriate place for a canal. Parallel alignments of the main canals, maintained over long distances, to his discerning contrasted sharply to what he called "the tangled meaningless mass of waterways...where we nature's undirected handiwork." Willcocks tried to investigate these dead rivers, locally called nadis (or blind rivers). He concluded that during the flood season, the embankments along) artificially constructed canals would be regularly breached to take the floodwaters to the fields the British administrators misunderstood them for flood embankments. And considered the re breaches in them as discreditable occurrings.

Therefore, the government did not do anything to i these canals and maintain them. Even worse, wherever possible, it constructed solid embankn prevent wholesale breaches. "But," says Willcocks, "it ought to have been evident that a breaches in an heavily embanked river of inconsiderable length could not possibly have been \ by the river itself." Willcocks further argued that "riverwater in the early months of the floo gold," because it is full of extremely fertile mud and fish eggs.

The resulting destruction o overflow irrigation system of Bengal steadily lead to a decline in agricultural production, increase malaria and the famous famines. Willcocks cities the specific example of how several were raised around the Damodar river and earlier ones strengthened "like Satanic chains"-; which were earlier regularly breached-and thus "doomed the once healthy and prosperous between the Damodar and the Hooghly which started facing the problems of malaria and poverty the 1930s, eminent economist Radhakamal Mukherjee, described the lack of flush Bengal, particularly in Burdwan, Hooghly and parts of Bankura and Midnapore, as a "Red Famine."

In very dry areas, the primary need of the people is drinking water, and various ingenious were developed to collect rainwater for use as drinking water. In Rajasthan, for instance, there old tradition of using rooftops as a catchment area to collect rainwater. In the town of Phalodi, i household does not want to collect water from its rooftop, then neighbours borrow the roof to o water for their needs. And, in areas where land is not a limiting factor, people have even developed customised rain harvesting structures called kundis to obtain drinking water. These are amazing water harvesting structures which dot the landscape of the district of the Thar desert. Such structures have been reported from any other part of India or the world. Kundis are artificial wells which store from an artificially prepared catchment surrounding the well so that any rainwater that falls 01 catchment rapidly runs into the well and gets stored. Kundis can be made anywhere if adequate is available, of which there is no shortage in the desert. Its potential can be understood following fact. If an area receives only 100 mm. of rainfall-which would make it an extremely environment-this rain harvested over one hectare of land would provide one million litres of wa year. Kundis are common in the Thar desert of Rajasthan, and they teach us the amazing lesson the land, wherever it may be, can be used to make a kundi and harvest rainwater. As a family o would not need more than 10-15 litres a day for drinking and cooking or 3650-5475 litres a 180-270 families could meet their most critical water needs by making a one hectare kundi. The nomadic Maldharis of the inhospitable Kutch region of Gujarat have developed an' interesting system of procuring potable sweetwater, even in an area where rainwater is scares ground-water is saline.

Hence, it is theoretically possible to keep the harvested sweet rainwater stored in a way that i continue to float on the denser saline water and, thus, provide the people with an opportunity and survive in such a hostile region. They have, in fact, through their own experience and know developed precisely such a system, locally called virda, which is essentially like a well in a tank.

In terms of carrying water over a difficult terrain, the indigenous people of northeastern India, their traditional expertise in the use of bamboo, have developed some of the most interesting artistic systems. All over the eastern Himalaya and the northeastern hill ranges, people build bamboo pipelines to carry water from natural springs to a convenient point, where it Cc used for drinking. But this art has been raised to a sublime level by the people of soul.

Meghalaya, who live near the border of Bangladesh and have used intricate networks of bar pipelines to deliver water to betel leaf plantations in rocky areas, where no channels can be built entire system literally works like a modern drip irrigation system which delivers measured of water straight to the roots of the plants. Some 18-20 litres of water enter the bamboo systems every minute and after getting transported over several hundred metres, finally reduced to 20-80 drops per minute at the site of the plant.

Thus, whether it is rainwater, streamwater, floodwater or ground-water, people have develop ingenious ways to harvest and deliver that water for irrigation and for human and consumption.

Systems of Management

Water harvesting systems are fragile entities. They have to be continuously maintained and even more difficult is to apportion the scarce water amongst the farmers. But across the people developed a variety of different property rights systems to share the water and to the systems. To ensure equity in the distribution of the scarce water, villagers in Ladakh elect a water official < year known as churpun, at the start of each agricultural season. The churpun ensures that farmer gets adequate water in proportion to the area of his land and that no field is left unirrig.

Hence, disputes over use of water are very rare. Canals are repaired by community effort. Position of churpun rotates amongst all households to ensure that no household corners this ci position.

In order to share the water for their terraced rice plots, villagers in Nagaland have also develop variety of practices. While water sources are common property, land is private property. Then cases where water is taken over channels several kilometres long to reach the fields. In some villages, local chiefs will stand on a rock on a particular day in the months of May, June and July ask the villagers to offer prayers for good water and clean the traditional water sources.

Angamis, Chakhesangs and Zeiiang tribes, traditional water rights are very rigidly held. These \ rights may be held by individuals, clans or khels, or villages. In Khonoma village of Kohima for instance, the person whose field is closest to the water source is responsible for conservim source. But the person whose field is at the bottom of the canal is its owner, he has a major she water and he also regulates the flow of water. He is also responsible for keeping the canal c

While every one has the right to take water from the stream, nobody can take more demarcated share.

In Kikruma, a Chakhesang village located in the Phek district, a combination of water rights exist a spring exists on a particular terrace which is not yet cultivated, the person below it is use it to develop his terrace. But in case the owner of the uncultivated terrace decides to develop terrace at some later date, then the two will share the water equally. In general, the rights runoff between different terraces is respected.

The maintenance of all channels catchment areas of ponds is done once every year, usually before the onset of the monsoon, responsibility of mobilising people to clean the channels rests with the last person tappinc channel, locally referred to as the Neipu, which literally means the lord. The community-managed phad irrigation system, prevalent in Dhule and Nashik district northwestern Maharashtra, probably came into existence some 300 to 400 years ago. A bandharas (dams) were built on the local rivers to divert water. In the phad system, variatio water availability from one year to another were managed by demarcating the command area two categories: one with assured and another with unassured irrigation. In years of good rain unassured area, which is invariably at the tail-end, also gets irrigation benefits. Only one crop Cc grown in one phad and the variety of crops to be grown in different phads is decided by an irrigators, depending on water availability. Crops are rotated among the phads so that o' period of two to three years, every phad gets to grow sugarcane, the main cash crop. In D.

village, for instance, the total cultivated land is divided into four phads. Every landowner has a p land in each of the four phads. In a year of plentiful water, the community may decide to sugarcane in three and millet in one. In a year of average rain, the farmers may grow tow sugarcane and two of millet. In a bad year, the community may allow sugarcane in only one, millet in two and even keep one fallow.

In Tamil Nadu too, which has a profusion of tanks, local irrigation institutions have evolve develop a common code for maintenance, water sharing and resolution of conflicts. They control all activities relating to tanks, including the maintenance of the water channel, tank bed, strengthening of the tank bunds, maintenance of the regulatory structure and channi the command areas, and regulation of water supply. The organisational structure operates, first a supervisory level as an enforcing authority, and, secondly, at a more menial level involving labour.

The first level of officials is called nattamaikar (a group of persons representing the e village) or kavaimaniyam (generally one person). While the kavaimaniyam is an irrigation concerned exclusively with organising maintenance work, nattamaikar organise irrigation-re activities as well as other village activities such as temple festivals. Nattamai and kavaimaniyar assisted by irrigation workers called neerkatti, whose major responsibilities are to inspect the channels during monsoon months, and to inform the nattamaikar or kavaimaniyam about breach illegal tapping of water by upstream farmers, watch the bunds and surplus weirs particularly d floods, irrigate lands as per the directions of the kavaimaniyam or nattamaikar, and protect the from cattle.

The number of neerkattis varies from one to 10, depending upon the size o command area. They are paid in kind every season by the farmers of the concerned tank Generally, wages are pooled on the basis of command area cultivated, and shared equally among irrigation workers. Nattamai or kavaimaniyam generally call a meeting of the landholders in the command area as a tank starts receiving water through its feeder channels. Neerkattis announce the beating drums.

The meeting decides when to start the maintenance work and the contribution labour by each landholder. If a landholder fails to contribute his labour then he must substitute hired labour or else he is fined. In some tanks, the kavaimaniyam maintains a record on the labour involved in maintenance work. To avoid excess water being drawn by a channel at its head, thereby leaving less water downstream users, a simple device has been adopted by the Central Himalayan villages of Pradesh. At the mouth of each turnout, a small boulder is placed. The stone size at successive \ turnouts gets progressively smaller. Some measure of equity is thus ensured in water distribution Water distribution in Ladyura village, for instance, is carried out by an irrigation committee which 10 members, including a chowkidar (guard). Many non-elected members who are familiar with \ distribution arrangements are co-opted as members. Major decisions are taken in a general meeting which is open to all residents of the village. There is a mechanism to resolve disputes, L which the irrigation committee members of the two upstream and downstream villages sharini same channel sit together to discuss the matter. Records of attendance and minutes of meeting meticulously recorded. Considerable interest is taken by villagers and committee members proper running of the committee and the irrigation system. The committee declares the irrigation and canal clearance repair work. The Role of the State in Promoting Water Harvesting Systems-Pre-colonial Rulers.

When the British came to India, they saw a land extremely rich, highly urbanised and literate, and where arts, crafts and literature flourished. The wealth of the land came from its resource mobilisation. The surplus generated in the villages not only supported the villages but the cities and towns. Over centuries, Indians had learnt to use their land-water-vegetation in an intelligent and sustainable manner. The resources base around each village had transformed into a complex agroecosystem of croplands, grazing lands (grasslands), and foresi tree lands, thus constituting an interactive multi-component biological system that respondec only to the region's sharp seasonal rhythms but also reduced risk by keeping the social and occur impact of rainfall variations down to a minimum.

Due to the seasonal nature of rains, they catch the rainwater that fell in their own village or the streamwater that flowed past them during monsoon season. When the British came, there were already hundreds of thousands of tanks a the country. The Rule was Fiscal Incentives not Bureaucracies and Let the People Do it Themselves The Indian rulers preceding the British did not have irrigation bureaucracies or public v departments to create these structures. In the Thar desert of Rajasthan, says traditional knowledge expert Komal Kothari, "while collecting information from some 600 villages, I found that the c the jagirdar or anybody who had anything to do with revenue collection, did not create any \ body for the people.

All water bodies made by the erstwhile kings, jagirdars, chiefs and chief were reserved for their personal use. The people largely had to fend for themselves." Though the of the state varied from one region to another, it was nonetheless true that the rulers rarely irrigation structures themselves. The massive Pichola lake in the famous city of Udaipur, for inst. was built by nomadic gypsies. But the rulers played an important role in encouraging people to water harvesting structures. The famous Vijayanagar kings of South India, for instance, placed great importance on develop irrigation facilities for agricultural important.

The officers of the kingdom, private individuals public bodies in the Vijayanagar empire constructed tanks. In those days, land tax, which collected in kind in the form of one-sixth of the produce, was an important source of revenue for Indian kings. Thus, if agricultural production increased kings became richer and if it dropped, too became poorer like the populace. The state, therefore, had a vested interest in encourage; private initiative to develop irrigation systems. The Vijayanagar kings would make kattu kodage grants to individuals or institutions that undertook such works. The enterprising p< would be granted a piece of tax free land watered by the tank, canal or well which he/she the extent of the grant varied with the importance of the work. Thus, when one Harinideva constructed a tank in Mysore district he was given a grant by Emperor Deva Raya II (1423-1@ and, when the tank was extended by him, another grant was made to him. Everybody Encouraged Water Harvesting Including Temples Dasavanda grants were made not just by the kings. A record of 1497 in the present Chittoor di registers a Kattu kodage grant of land at Gundalahalli with certain stipulations about the produce. The gift was made by the sthanika (manager) of a temple to one Narasimhadeva for a tank in the village belonging to the Kadiri Laksminarsimha temple and for bringing the surrout lands under cultivation.

Great religious merit was also attached to the maintenance and repair of irrigation works Vijayanagar kings. An inscription of 1413 states: "A ruined family, a breached tank or pond, a 1 kingdom, whomsoever restores, or repairs a damaged temple, acquires merit four-fold of that \ accrued from them at first." Temples too played a pivotal role in irrigation management. The famous temple of Kalahasti, Tirupati in Andhra Pradesh, for instance, used devotee endowments to excavate irrigation and to reclaim temple lands.

A Kalahasti inscription of 1540 states: "Virappanar Ayyar (with the treasury of the gods at Kalahasti) for the purpose, 1,306 pond, which was to be invest the new settlements of Muttayammansamudram...with a view to bring the lands under cultiv (and) the lands of Lakkusetipuram to be irrigated and brought under the plough with the help c waters of Virasamudaram lake (which will) be repaired and maintained by investing 1,006 pond amount deposited." It made good economic sense for the temples to invest in irrigation. Gett "regular share in the production from the irrigated lands meant that they then had a regulatory.

Temples were sometime also granted tax-free land by the state and temple officials then le lands for reclamation and cultivation. Water Harvesting in the Gond Empire The tribal Gond kings were great empire builders. By the 9th century AD, the whole of the ea section of the Central Provinces had come under Gond influence and was known as Gondwana (of the Gonds). Gondwanaland, this famous geological continent had a practise similar Vijayanagar kings.

The pre-colonial states of Patna and Sambalpur (now in the state of Orissa have ruins of the old irrigation works. Building of reservoirs for irrigation was the foremost the village chief. Though the Gonds had a strong central government, each village was independent in its economy and governance. The village was primarily a settlement of peasants, and its an association of cultivators. Its prosperity rested on proper management of land and \ resources. Repairs of channels, embankments and distributaries were immediately taken up after first rainfall. Negligence invariably came in for scathing criticism by other village. The Gond gave to anyone who made a tank, a grant of revenue-free land lying below it. Agricultural increased under the Gond rulers, and great works like the Rani talab near Jubalpur remain to thi: a monument of their rule.

Rent-free land was given to Kodas, a class of people who were excavating water reservoirs, and looked after the maintenance of irrigation works. Such land g were known as sagar rakshya jagit. The Decline Began With the British British rule, unfortunately, laid waste to this enormous heritage. In their desire to rule, admire and maximise their revenue from this rich land, the British steadily impoverished the communities and led to the destruction of their resource management systems, including the \ management structures that had emerged over the centuries. Early British observers like Ch Metcaife and I Ienry Maine had described India as a largely happy land of "village republics'.

What meant, says historian Dharampal, is "that the 'village', to an extent, had all the semblance o State, it controlled revenue and that it exercised authority within its sphere....The basic element this 'village republic' was the authority it wielded, the resources it controlled, and dispensed, an manner of such resource utilization...Indian society and polity had basically been organised according to non-centralist concepts...That the annual exchequer receipts of the Mughal Emperor Jahangi not amount to more than 5 per cent of the computed revenue of his empire, and that of I Emperor) Aurangzeb with all his zeal for maximising such receipts, did not even exceed 20 per c< symptomatic of the concepts and arrangements which governed Indian Polity...there is volume data scattered in the British records themselves which confirm the view, that in terms of the expenses, both education and medical care, the expenses of the local police, and the maintaining irrigation facilities, had primary claims on revenue..." Dharampal points out that the data collected by the British in the 1770s and 1780s from Bengal, Bihar showed that revenues of these areas were divided into several categories.

The 'khalsa' those sources whose revenue was received in the exchequer of the ruling authority of the provision Two of the other categories which probably accounted for about 80 per cent of the computed rev of an area were termed 'Chakeran Zemin' and 'Bazee Zemin'. The former implied such recipes revenue who were engaged in administrative, economic and accounting activities and remunerated by assignments of revenue. The latter meant those individuals, groups and institute who, according to the British, were in receipt of what were termed "religious allowances". Assignments under 'Bazee Zemin' were quite high. In Bengal, the British noted "almost one half of the province is held upon free tenure" under the 'Bazee Zemin' category. In t districts of Bengal and Bihar there were as many as 30,000 to 36000 recipients per districts. After the turmoil of 1750-1800, the period during which the British took control of south India situation, as recorded by the British, in the Madras Presidency was not much different. As la 1801, over 35 per cent of the total cultivated land in the Ceded districts (the present Rayalsi area and the district of Bellary) came under the category of revenue-free assignments. The which constituted Malabar had, till about 1750, never had a land tax, says Dharampal, "What true of Bengal, Bihar and the Madras Presidency applied equally to other areas in India, whett-the areas of the Bombay Presidency, or of Punjab, or in the Rajasthan states."

By about 1800, a very large percentageof these revenue assignments had been dispossessed, reducing their beneficiaries to penury and beggary, and most of the remaining their assignments greatly reduced so that they could no longer perform the educational, \ management or other functions that they were expected to undertake. It was the task of Thi Munro to reduce revenue-free assignments in the Ceded districts to a mere 5 per cent of the cultivated land. The revenue thus collected was drained out of the land and out of the leaving behind a devastated natural resource management system.

With the destruction o indigenous financial system, community property slowly became nobody's property. Furthermore, the colonial rulers neglected what they could not own or that which did not monetary benefits. As Arthur Cotton, the pioneer of modern irrigation in India, noted in 1874 c local water harvesting systems: "There are multitudes of old native works in various par India....These are noble works, and show both boldness and engineering talent.

They have stood hundreds of years. When I first arrived in India, the contempt with which the natives justly spot us on account of this neglect of material improvements was very striking; they used to say we kind of civilised savages, wonderful expert about fighting but so inferior to their great men, the would not even keep in repair the works they had constructed, much less even imitate the extending the system." When the British realised their mistake, they tried to take remedial measures but totally faith understand the strength of the indigenous system.

As a result, the measures they undertook failed. Firstly, they created irrigation and public works bureaucracies which were supposed to and manage the neglected water harvesting systems. When these technocracies failed to decentralised systems, the British authorities in the Madras Presidency, in the mistaken belief, the local communities would as earlier undertake voluntary labour to maintain the tanks tradition, enacted the notorious Madras Compulsory Labour Act of 1858, which created consider resentment amongst the peasantry. In Bengal, as William Willcocks pointed out, the British failed to understand the technological nature of the indigenous irrigation system. With the progress of British rule, there was a gradual a shift in emphasis from minor irrigation v like tanks, wells, bandharas and small river channels to large dams and canals extensive areas. The latter were works and little importance was given, in comparison, to the works. Indian rulers, on the other hand, continued to build and maintain tanks in states not rule the British. While the British were Destroying Tanks, the Nizam in Neighbouring Hyderabad Building Them.

In contrast to the erstwhile Madras Presidency area, there is evidence that tank irrigation former Hyderabad state was of more recent origin. The area irrigated by tanks increased consider during the first half of the 20th century under the rule of the Nizams of Hyderabad. From 4,500 !@ 1895-96 the Public Works Department (PWD) of the Hyderabad state reached around 45,000 tank irrigated area around the turn of the century, and between 324,000 ha. and 364,500 ha. : 40 years later. The fact that the PWD of the Hyderabad state expanded the tank irrigated area the British government in the Madras Presidency did not do so, indicates that the intensity of irrigation in different areas was greatly influenced by the policies of their respective governments.

Even worse, the British, in their bid to rule, educated an entire class of Indians which no appreciated or understood India. They were, in fact, so successful that when India being independent, the leaders of modern India also turned their back on these systems. Indians almost exclusively in mega-irrigation systems greatly influenced by the technological socialist countries like the Soviet Union. Over time, other changes have taken place in several of India which have further eroded the traditional water harvesting systems.

Overall, there has erosion in community self-management as bureaucratic intervention in village affairs has steadily encouraged by India's political leaders. Technological changes like the interdictive tube-wells means that richer farmers in the command area of a tank who can install these tube no longer have a vested interest in cooperating with the rest of the community in managing tanks. The future of the phad system in Maharashtra too is now uncertain.

Firstly, because government unthinkingly built reservoirs upstream to utilise the available river waters, thus affect the post-monsoon flows downstream, where the bandharas (dams) feeding the phads with \ were built. Secondly, a sugar factory has come up in the area, increasing the demand for sugan a water intensive crop. Urban Tanks are in a very Bad State Many central and southern Indian cities like Hyderabad, Madras and Bangalore grew up traditional water harvesting systems. But in all these urban areas too, the systems have e disappeared because of pressure from real estate lobbies or they have become heavily polluted.

Result, traditional water harvesting systems continue to play an important role largely in re areas where the reach of water bureaucracies remains weak as in many Himalayan states. What Exists ? The first census of minor surface irrigation works conducted in 1986-87 revealed that some 62; works were still in use with an irrigation potential of 7.6 million hectare, though they were in a dilapidated state. Some 53,000 works had gone totally out of use. Those still in use included 14{: tanks, 95,000 stream diversion works and 350,000 surface lift irrigation schemes. The Potential But the potential of these water harvesting systems is once again being recognised and del because the large water supply systems built around mega-dams have proved to be extra capital-intensive, their gestation periods are long, and their returns in terms of increased productivity has been rather low, largely because the water supplied by these systems is deterring more by bureaucracies controlling these systems rather than the farmers.

Irrigation efficient' tube-wells, because they are farmer controlled, has been much better but the rapid increase: tube-wells in dry areas, without the supplemental ground-water recharge that used to be ensure traditional water harvesting systems, is leading to a rapid depletion of ground-water aquifers an' become a severe problem in the years to come. Rain water harvesting systems, on the other hand, require small sums of money, a large pc which can come from local communities and thus avoid an undue drain on the state exchequer; can be built within months instead of years like large dams; they will be under the control o farmers themselves; and, they can be used conjunctively with tube-wells as they can rechargi ground-water aquifers. In terms of the water they can store, their potential is, of course, stupendous.

If 5 per cent of land area, about 15 million hectares, was used to store water at an average depth of 5 metres, would be able to get 37.5-75 million hectare-metres of water, depending on the rainwater colle efficiency ranging from 50 per cent to 100 per cent. The potential of rainwater harvesting to supply clean drinking water is even more stupendous. 1 is theoretically speaking no village in India which cannot meet its drinking and cooking water through rainwater harvesting.

If an individual on an average needs 2.5 litres of water per cooking and drinking, a billion people, India's expected population by the turn of the century need about 2.5 billion litres per day or 912.5 billion litres per year. If rain was harvested over a 83,000 hectares or 830 sq km every year, clean drinking and cooking water could be obtained country's entire population. And if the nationwide supply of water was stipulated at 100 litres per person per day, the norm in many urban areas, then too all that we would need is 1 per cent of India's land area to aside for rainwater harvesting.

BASIC CONFIGURATION

Rainwater harvesting systems channel rainwater that falls on to a roof into storage via a system of gutters and pipes. The first flush of rainwater after a dry season should be allowed to run to waste as it will be contaminated with dust, bird droppings etc. Roof gutters should have sufficient incline to avoid standing water. They must be strong enough, and large enough to carry peak flows. Storage tanks should be covered to prevent mosquito breeding and to reduce evaporation losses, contamination and algal growth. Rainwater harvesting systems require regular maintenance and cleaning to keep the system hygienic.

Subsurface Dyke

A subsurface dyke is built in an aquifer to obstruct the natural flow of ground-water, thereby raising the ground-water level and increasing the amount of water stored in the aquifer.

The subsurface dyke at Krishi Vigyan Kendra, Kannur under Kerala Agricultural University with the support of ICAR, has become an effective method for ground water conservation by means of rain water harvesting technologies. The sub-surface dyke has demonstrated that it is a feasible method for conserving and exploiting the ground-water resources of the Kerala state of India. The dyke is now the largest rainwater harvesting system in that region. It is very lowest cost but earn more gain by the use of rain water harvesting.

Ground-water Recharge

Rainwater may also be used for ground-water recharge, where the runoff on the ground is collected and allowed to be absorbed, adding to the ground-water. In the US, rooftop rainwater is collected and stored in sump.

Advantages in Urban Areas

Rainwater harvesting can (a) assure an independent water supply during water restrictions, that is though somewhat dependent on end use and maintenance, (b) usually of acceptable quality for household needs and (c) renewable at acceptable volumes despite forecast climate change (CSIRO, 2003). It produces beneficial externalities by reducing peak storm water runoff and processing costs. In municipalities with combined sewer systems, reducing storm runoff is especially important, because excess runoff during heavy storms leads to the discharge of raw sewage from outfalls when treatment plant capacity cannot handle the combined flow. RH systems are simple to install and operate. Running costs are negligible, and they provide water at the point of consumption. We are consuming this water for our basic needs.

Rain water harvesting law.— Who Owns the Rain? Check water rights and water right law very well before buying land in USA. Some states have water law in favor of land owners, and some states own all water rights, leaving you only water rights they are willing to grant through permits. In Colorado you may not catch, collect or harvest rain water from your roof unless you first buy a permit. Depending on where you live, and if the government owns your rain water rights.

Quality

As rainwater may be contaminated due to pollutants like microscopic germs etc., it is often not considered suitable for drinking without treatment. However, there are many examples of rainwater being used for all purposes — including drinking — following suitable treatment.

Rainwater harvested from roofs can contain human, animal and bird feces, mosses and lichens, windblown dust, particulates from urban pollution, pesticides, and inorganic ions from the sea (Ca, Mg, Na, K, Cl, SO_4), and dissolved gases (CO_2, NO_x, SO_x). High levels of pesticide have been found in rainwater in Europe with the highest concentrations occurring in the first rain immediately after a dry spell; the concentration of these and other contaminants are reduced significantly by diverting the initial flow of water to waste as described above. The water may need to be analysed properly, and used in a way appropriate to its safety. In the Gansu province for example, harvested rainwater is boiled in parabolic solar cookers before being used for drinking. In Brazil alum and chlorine is added to disinfect water before consumption. So-called "appropriate technology" methods, such as solar water disinfection, provide low-cost disinfection options for treatment of stored rainwater for drinking.

System Sizing

It is important that the system is sized to meet the water demand throughout the dry season. Generally speaking, the size of the storage tank should be big enough to meet the daily water requirement throughout the dry season. In addition, the size of the catchment area or roof should be large enough to fill the tank. Edava has the cleanest lake in the whole world.

Around the World

- Currently in China and Brazil, rooftop rainwater harvesting is being practiced for providing drinking water, domestic water, water for livestock, water for small irrigation and a way to replenish ground water levels. Gansu province in China and semi-arid north east Brazil have the largest rooftop rainwater harvesting projects ongoing.
- In Rajasthan, India rainwater harvesting has traditionally been practiced by the people of the Thar Desert.
- In Bermuda, the law requires all new construction to include rainwater harvesting adequate for the residents.
- The U.S. Virgin Islands have a similar law.
- In the Indus Valley Civilization, Elephanta Caves and Kanheri Caves in Mumbai rainwater harvesting alone has been used to supply in their water requirements.
- In Senegal and Guinea-Bissau, the houses of the Diola-people are frequently equipped with homebrew rainwater harvesters made from local, organic materials.
- In the United Kingdom water butts are often found in domestic gardens to collect rainwater which is then used to water the garden. However, the British government's Code For Sustainable Homes encourages fitting large underground tanks to new-build homes to collect rainwater for flushing toilets, washing clothes, watering the garden and washing cars. This reduces by 50% the amount of mains water used by the home.

- In the Irrawaddy Delta of Myanmar, the ground-water is saline and communities rely on mud-lined rainwater ponds to meet their drinking water needs throughout the dry season. Some of these ponds are centuries old and are treated with great reverence and respect.
- Until 2009 in Colorado, water rights laws almost completely restricted rainwater harvesting; a property owner who captured rainwater was deemed to be stealing it from those who have rights to take water from the watershed. Now, residential well owners that meet certain criteria may obtain a permit to install a *rooftop precipitation collection system* (SB 09-080). Up to 10 large scale pilot studies may also be permitted (HB 09-1129). The main factor in persuading the Colorado Legislature to change the law was a 2007 study that found that in an average year, 97% of the precipitation that fell in Douglas County, in the southern suburbs of Denver, never reached a stream—it was used by plants or evaporated on the ground. In New Mexico, rainwater catchment is mandatory for new dwellings in Santa Fe.
- Kerala, India,
- In Australia rainwater harvesting is typically used to supplement the reticulated mains supply. In south east Queensland, households that harvested rainwater doubled each year from 2005 to 2008, reaching 40% penetration at that time (White, 2009 (PhD)).

URBAN RAINWATER HARVESTING

As ground-water levels drop all over India, New Delhi regularly faces water shortages in the summer. A primary school teacher in New Delhi began a collaborative project to design and install rooftop rainwater harvesting, which has raised the local water table and dramatically reduced the school's dependence on the municipal water system. The Sri Ram school is part of a larger movement to install rainwater harvesting systems on buildings and apartment complexes around the city. Encouraged by its success and support from the community, the school has started numerous other environmental projects as well, such as a "zero-waste" policy, vermicomposting and no-plastic drives. From about 1997, water shortages began making headlines in India. Delhi, receiving only 611 mm of rain annually, would be one of the urban centers hard hit by droughts. Madhu Bhatnagar, at the time a social studies teacher at a public elementary school in New Delhi, began asking herself what could be done about the high level of water consumption in schools, though at the time she didn't know anything about rainwater harvesting (RWH). By chance, Bhatnagar attended a weekend seminar on rainwater harvesting hosted by CSE (the Center for

Science and Environment, a Delhi-based environmental research institute). In the summer of 1999, with money and the full support of the principal and the Parents' School Association (PSA), she teamed up with an engineer from CSE and a representative from the Central Ground-water Authority to design the system. It involved the laying of drains and pipes leading down from the roof where water is captured from three points on 6,000 square meters of rooftop. The water, which used to just drain into the sewers, was funneled down via a system of underground pipes which eventually drained into a filtration pit in the schoolyard to recharge the ground-water.

The overflow from this tank is channelled into a borewell 200 mm in diameter and 30 m deep. A single borewell supplies the whole school, and is supplemented by the municipal water supply. The borewell is connected to pipes which lead into the school's plumbing system, and the water for drinking goes into a separate filtration system. The total volume of rainwater harvested was 1,890,000 litres, or 55% of the total water harvesting potential.

The system cost Rs 125,000, or about US$3,000. It became operational in May 2000, and was inaugurated by the Chief Minister of Delhi as the capital's first school to harvest rainwater. At the inauguration ceremony held for the parents and children, some interested architects also attended. Parents who were building houses at the time also brought their builders, so that they could see the system and meet some of the people involved with designing it. In this way some parents began doing it in their homes and pressuring their residents associations to install rainwater harvesting structures in their colonies (apartment complexes). "It gathered momentum, now it's a household word," says Bhatnagar. CSE continues to monitor Shri Ram School's water table and quality as it does for all the RWH projects it supports.

Since the beginning, ground-water levels have risen from 45 meters below ground level in 1999 to 34 meters today, and dependence on the municipal water supply today is marginal. Word has spread quickly, Bhatnagar says, because school is "a learning point" with children as a powerful transmitter of ideas to parents and so are a crucial a link to the larger community.

"I feel the children are a source of power, [and] we are the starting point of campaigns which spread to other schools," she says. Programs like this have both spearheaded and ridden on the tide of a citywide movement to harvest and recycle water, and a municipal law will go into effect next year mandating all new buildings to harvest a certain amount of rainwater from roofs whose area exceeds 100 square meters, and on all plots of land with an area of 1,000 square meters. Other legislation comes from The Central Ground-water Authority (CGWA), making RWH mandatory in all institutions and residential colonies in notified areas, as well as in notified areas with borewells, to go into effect in March of 2002. Moreover, the CGWA has banned the drilling of borewells in notified areas. In the sprawling Vasant Vihar area where the school is located, there are 1,350 plots of land, most of which have borewells. The CGWA made RWH mandatory to those plots with borewells or they would be sealed over. This, along with examples like Shri Ram, have spurred residents into action, with 11 public projects and 30 more in the works.

The success of RWH has provided impetus for a variety of other school programs. Manika Sharma, the school principal for the past decade, gave teachers the freedom to put their ideas into the curriculum and programs. For Madhu Bhatnagar, many opportunities arose to teach kids about sustainability. For example, old traditions from India's multiple festivals which are celebrated in schools, have taken on new habits such as fireworks and plastic decorations.

Shri Ram took this as an opportunity to promote safe and sustainable festivals and campaign against the use of firecrackers. The school took a holistic approach to its campaigns. "It wasn't just about lighting the firecracker and air pollution, it also goes back to the production of the cracker, where does it come from, who's involved? Young children, safety of the children, child labor–all of that," Sharma explains. From there, the school had a no-plastics drive where cloth bags were made and given to parents so they would not bring plastics into the school. They also discouraged parents from covering the students notebooks with a plastic sheet as is common practice in other elementary schools. "The parents were very much in agreement with what we believed in, which helped a lot because it became a community project," Sharma says.

While the school was already doing innovative projects, creation of the RWH system was a springboard for programs to follow. An extensive "zero-waste zone" has been created, with all garbage separated at source in the classrooms, and large-scale paper recycling both in the office and the classrooms began. At the end of every year, there is a "book bank," where children return their old half-used notebooks. Some of the more colourful or interesting pages are taken out, made as covers, and the unused pages are rebound again as new books. Another aspect of waste recycling was the kitchen waste for composting.

Waste from the school lunches is composted on-site with two pits. One does vermicomposting, and the other uses enzymes. The compost is bagged and sold to parents, or it is used in the garden where vegetables such as eggplant and tomatoes, flowers and medicinals are grown organically. The school has also begun experimenting with CSE to begin greywater recycling in the school which would be used in the gardens. Most of the funding for these activities has come from the Parents' School Association which holds fundraisers, but recently Bhatnagar has been looking at other sources of money, for example from multinationals, to start more ambitious projects like getting solar panels or starting recycling programs in other schools.

Some challenges have been to get other schools to begin RWH. Replication of the system has had more success with parents in their colonies or homes than with the schools. None of the 14 neighbouring schools have started RWH yet. Sharma says the reason for this is financial. "A lot of schools back off to say, 'oh a couple of lakhs (one lakh is one hundred thousand) [rupees], 3-4 lakhs....we'd much rather buy chairs, or desks or whatever....so they don't think long-term," she says.

Bhatnagar has moved from teaching social studies to the design an environmental curriculum and promoting it through a network of 23 public schools. The curriculum is a 30-minute capsule once a week and all levels from the four-year-olds to the highest classes are part of it. This includes field trips and "yatras" (walks) to various important places, including a walk from the near the source of the Yamuna River to where it runs through to Delhi (as Bhatnagar says "from purity to profanity").

Shri Ram is the lead school for the 25 schools "eco-clubs" where children get together for various environment and nature activities. Other school-wide projects have been to campaign for tigers, where a group of children presented letters to the government to protect the endangered tigers in Sariska and Ranthambore sanctuaries. They also raise money for "ragpickers'" education (children who rummage garbage for material to sell), and using recycled materials at festivals.

The campaigning has not always made Bhatnagar popular. For example, stopping firecrackers or burning of dieties at festivals has provoked some backlash in the community that the school is interfering with traditions, and this period, Bhatnagar admits, was "tough." But she says the parents have told her how much more aware and motivated their children have become. "They say, 'Mom, just look at that house, there the water tank is leaking, so why don't we go and tell them?'...they jot down names of polluting vehicles and they stop people from littering on the road," she explains. "Kids are great activists."

INDIA - EAST CALCUTTA-MAKING THE MOST OF IT: WASTEWATER, FISHPONDS, AND AGRICULTURE

The East Kolkata Wetlands is a highly complex, and poorly understood, system. It processes the city's waste—both garbage and sewage—a service that would be very expensive to replace. It is also home to numerous people and animals. This case demonstrates the important EcoTipping Points principle of "nature doing the work."

Sometimes called the "kidneys of Kolkata (Calcutta)," the East Kolkata Wetlands (EKWs) are the largest of their kind in the world, covering an area of 12,500 hectares which were designated as a RAMSAR* site in 2002. This multiple-use wetland lies east of the city and includes a garbage dump known as Dhapa Square Mile, a mosaic of vegetable fields, a series of 300-odd fishponds connected by major and secondary canals, rice paddies, wholesale markets, a few roads, and 43 villages (with 60,000 people in all).

The EKWs extend almost equally on both sides of a Dry Weather Flow Channel, which discharges into the Kulti Gong (the wastewater outfall of Kolkata 28 square kilometres to the east). The area is divided into eleven zones and includes four major sub-regions: freshwater fishponds, brackish fishponds (both of which are known as bheris), garbage farms, and paddy lands. Some thirty kilometres eastward, the river Kulti-Bidyadhari drains into the Bay of Bengal. Underneath the city is a large supply of groundwater. The fishponds produce some 13,000 tons of fish annually, whose yield at 2-4 times higher than average fish ponds, is among the best of any freshwater pisciculture in the country. Some 150 tons of vegetables per day are harvested from small-scale plots irrigated with wastewater.

Geography

The EKWs are part of a region which is the mature delta of the Ganges River where its tributaries eventually drain into the Bay of Bengal. The climate is hot, humid and monsoonal, with the average yearly rainfall at about 1600 mm (mainly between mid-June to mid-October). January is the coolest month while May is the hottest. The tropical region is naturally suitable for using solar radiation to improve wastewater quality.

The EKWs are home to over 100 plant species, rare mammals including the marsh mongoose, the small Indian mongoose, the palm and Indian civets, and the threatened Indian mud turtle. There are also some 40 species of birds, both local and migratory, which include kingfishers, grebes, cormorants, egrets, terns, eagles and sandpipers. The wetlands act as a "sink" for the city of Kolkata, which has no conventional sewage treatment facility for its 12 million inhabitants. Some 3,500 tons of municipal waste, and 680 million litres of raw sewage enter the wetland system every day. Still, only 30% of the total wastewater is used for aquaculture or irrigation, while the remaining 70% flows directly into the Bay of Bengal (which pollutes the estuary, highlighting the need for improved efficiency of the system).

Land Use History

Throughout the ages, urban wetlands, either natural or manmade, have been a part of many Indian cities, which were preserved by people as the main source of water supply for drinking and irrigation. From the 15th century, the Ganges changed its main flow from the Bhagirathi to the Padma River, and this eastward change in direction profoundly altered the process of delta creation in south and central Bengal state. A number of tributaries were cut off from any upland flow and that signalled the end of these channels. But human interference in the region further cut off this spill area and channel beds rose, escalating the process of decay, which caused a river which fed into the wetland areas, the Bidyadhari, to dry up, stopping the silt deposition in spill areas.

Kolkata grew up along the Hooghly River, a tributary of the Ganges. The Hooghly River lies to the east and the now defunct Bidyahari River to the west. Since the mid-1800s, the area east of the city had been used as a garbage dump, where nearly 60% of its waste is delivered. When garbage was first brought to the area, the fertility of the soil began to improve (because the garbage had a much higher proportion of biodegradeable waste in contrast to the higher percentage of synthetic and nonbiodegradable content today).

The soil then became ideal for farming. At the same time, the drying up of the Bidyadhari spill channel caused the delta to deteriorate into a derelict, brackish swamp. In the 1930s, the Bidyadhari only carried city sewage and in the process became choked further because of the high silt content of the sewage.

When a Dry Weather Flow channel was constructed in 1943-44 to carry city sewage to the Kulti Gong (another river in the area), more wastewater was brought there, which increased its freshwater content. A large fish producer began growing fish in a large water area using city sewage. Local farmers stocked some of the ponds and dug new ones as well. Landlords, many of them absentee, leased most of the ponds to commercial managers, while others were managed by the government, and still others were given to fishermen's groups and co-operatives. These sewage-fed fisheries spread quickly, and were innovated, developed, and upgraded by local fish producers and farmers.

Part traditional knowledge, part enterprise and creativity, much of the folk technology is retained in the informal, oral tradition. The system reached peak productivity in the middle of the century but is now declining due to enocroachment and development fed by a real estate boom starting from the end of the 1980s.

Functioning of the System

In India, night soil and wastewater use in agriculture is a traditional practice which had been followed for centuries, but there are no other examples in India or the world of this scale and ingenuity. The EKWs are a delicate, complex and poorly-understood– and so highly vulnerable–system in danger of being swallowed by development. From the 1980s, the system was the focus of study by various academics, but this has failed to yield concrete results in protecting the wetlands. The West Bengal Government's Institute of Wetland Management and Ecological Design has also initiated various projects to study and conserve the wetlands in a practical process that includes participation by all those who depend on the wetlands.

It has worked with the UK's Department of International Development (the government's foreign aid agency) to create an action plan. Despite this, aspects of the system, such as the health issues of sewage-fed fisheries and contaminants in the garbage-raised vegetable plots, need further study. At the eastern edge of the city lies the fields which grow vegetables on the waste from Dhapa, the garbage dump. The vegetable production is a household activity, with people renting small plots or subletting smaller plots for household sustenance and income. These are designed with alternate bands of garbage-filled lands and long trench-like ponds known as "jheels," where sewage is detained for some time, and then used to irrigate vegetable fields.

The wastewater enters the wetlands through a network of drainage channels which flow into the canals and feed the fish ponds. On the way, the sunlight acts as a purifying agent on the sewage, which triggers biochemical reactions. For example, BOD (biochemical oxygen demand) is reduced through a symbiosis between algae and bacteria, where energy is drawn from algal photosynthesis.

Each hectare of a shallow water body can remove about 237 kg of BOD per day. This helps in the reduction of coliform bacteria prone to be pathogenic which even conventional mechanical sewage treatment plants may not be able to fully eliminate. The effluent from the fishponds is then made to drain further southeast where the paddy fields have been strategically located to benefit from the use of the effluent. Most of the fishponds range in size between two and ten hectares, but some individual ponds are over 70 hectares.

Three types of ponds are needed according to the stage of production: the nursery pond, the rearing pond and the stocking pond. Each needs a proper inlet-outlet management of sewage. The main requirement for a productive fish pond is the proper supply and quality of wastewater. Poor quality sewage reduces quality of nutrients, a higher toxic load on fish, and requires external inputs of nutrients. Countering this situation is done by allowing fish to grow bigger, but then this leads to a conflict of interests with unions, because this means a reduced number of harvesting days (and therefore days of work). The flow of water is mostly directed by gravity but in some areas diesel-powered pumps are used.

Fish are raised in five major phases: Pond preparation (done in the coolest months when ponds are drained, and maintenance or repairs of dikes are carried out), primary fertilization (initial introductions of wastewater into the pond and allowed to undergo natural purification, and stirring of the pond in order to reduce anerobic conditions in the sediments), fish stocking (where farmers initially stock a small number of fish to test for water quality, subsequently stocking up to four times), secondary fertilization (periodic introductions of wastewater into the ponds throughout the growth cycle), and finally fish harvesting (taken at different times according to species).

About 12 different species of fish are raised which occupy different ecological niches of the pond ecosystem, including Indian major, minor and exotic carp, varieties of non-native tilapia, mourala, and the freshwater giant prawn. In general, fish yields in the wetlands are found to be 2-4 times higher than those from ordinary fish ponds. Fish is generally sold in the fish markets in wetland areas or in Kolkata.

"Silt traps," which are three meters wide and 30-40 centimeters deep, are pits at the edges of bheris, which trap some of the silt buildup in the ponds themselves. These are periodically dredged and used to strengthen pond dikes. Ponds themselves are periodically drained before the monsoon season (which will naturally help to refill them) for various reasons: Draining them helps to free up some of the nutrients in the sediments if they are exposed to air, and it also kills some of the parasites that affect fish production.

Those who work in the wetlands either live on-site in the villages, migrate to various parts of the site depending on seasonal needs, or migrate in for work from outside. Jobs include farming, garbage sorting, fish farming, trading, auctioneering, selling, raising fish seed, making nets, maintaining drainage, and reinforcing banks. Many people depend on the wetlands not only for livelihood but also for the fish and vegetables produced in the area.

In "Wastewater-Fed Aquaculture in the Wetlands of Calcutta-An Overview," Dhrubajyoti Ghosh writes that there is an "outstanding demonstration of human skill in managing a productive system."

Water hyacinths also play a role in the system's function. By breaking up surface waves, erosion of the ponds' banks are reduced, replacing more expensive alternatives like stone or concrete to strengthen the banks.

Also, they provide shade to fish during the summer months, and the roots, well known to absorb metal ions, help to leach heavy metals out of the water. Occasionally they must be harvested, and they are used as buffalo feed or decomposed and fed to carp. Other important species in the system include domesticated animals such as cows, buffalo, pigs and ducks. Pigs run semi-wild and feed on whatever they can find. Ducks eat aquatic snails harvested from the ponds. Algae also plays an important role, as it not only provides food to fish, but supports certain bacteria which act on organic matter in the sewage.

Land Ownership

There are various types of land ownership of fish ponds:

1. Owner-managed (but this is disappearing because of uncertainties of the wetlands' future)
2. Cooperatives: Some 13 Cooperatives with about 300 farmers, these function well without any external assistance
3. State government corporation (only two ponds are under this control).

Much of the land is also leased out by absentee landowners, some of which are further sublet.

Income and Expenditure

This has not been studied extensively, and the reliability of existing data is poor, as private farmers may not disclose everything to researchers, and the State-managed fishponds are too few to fully represent the entire area.

Issues

Despite their potential to provide multiple natural and social services, the wetlands are an informal system. As such, they need further study, effective maintenance, monitoring, and upgrading.

Urban development: With encroaching eastward development of the city, there is speculation that certain areas will succumb to development. Whether or not this happens, this speculation alone is a threat to the future of the wetlands as it makes locals wary of investing their labor in long-term maintenance and upkeep of the ponds.

Pollution: Some unregulated industries, such as tanneries, have been releasing untreated effluent directly into the wetlands, further threatening the water quality.

Siltation of the ponds: Through natural processes, the buildup of silt has reduced the capacity of fish ponds over time. Some, originally six feet deep, have silted up to two feet from the surface, which has reduced fish production by two-thirds. The dredging and transport needed to clean out the larger ponds requires a large investment, even though the silt can be used for landfills planned for new townships north of the city.

Complexity: Because this is an urban area, there are more competing, conflicting interests and agencies than would be found in a village. Moreover, unlike rural development, the issues are less black-and-white, for example, income opportunities exist locally whether or not the wetlands are preserved (although the income and quality of life for those who depend on the wetlands might change dramatically if they were to disappear).

The presence of the Mafia: The presence of the Mafia in the control of fish ponds is a subject no one knew much about (or were not eager to talk about).

Safety of fish and farm produce: Although a study done by Ghosh et al. in 1980 found that the production of fish was not affected even by levels of ammoniac-nitrogen concentration of 5.13 milligrams/liter (the maximum limit is 0.1 mg/liter), there is a need to study the health issues of raising fish in sewage, and feeding water hyacinths, which may contain heavy metals, to water buffalo and fish.

The separation of garbage: While many Indian municipalities rely on the informal sector to supplement overburdened waste management services, there remain questions about the system's effectiveness and more importantly, the environmental and moral issues of relying on so-called "ragpickers" to sort municipal garbage. First, these are the poorest members of society, typically women and children. The work is filthy and dangerous as it puts workers—most of whom do not wear protective clothing or gloves—into contact with broken glass and medical and other hazardous wastes.

Many suffer from injuries and chronic skin diseases. Secondly, only garbage which can be sold is removed, which means that some non-biodegradable wastes that have no commercial value, such as plastic and foil wrappings, remain in the soil where vegetables are grown. (On our tour we saw the fields where vegetables were grown and it was full of tiny bits of foil and plastic.

We also saw piles of plastic garbage being burned, releasing foul-smelling black smoke into the air.) This brings into question the safety of the food and fish grown in the wetlands, with the possibility of contamination of soil or vegetables with heavy metals or other toxins. Third, there is the fact that a composting plant which has opened at Dhapa is taking biodegradeable garbage (which is sold to tea gardens) away from the vegetable farmers, so the soil quality is declining. As a result, vegetable farmers have begun using chemical fertilizers to replace the lost soil fertility.

Efforts to Preserve the Wetlands

Some NGOs have been lobbying for the preservation of the wetlands in fear that the "invisible" services this vast region provides—such as wastewater treatment, collection of rainwater run-off, preservation of biodiversity, food production, and income generation—will be lost, resulting in an increased financial burden to replace them. There have been some efforts to address some of the wetlands' problems by various government agencies under their jurisdiction, including the Department of Fisheries, Department of Agriculture, Irrigation and Waterways, the Kolkata Municipal Corporation, the Kolkata Metropolitan Development Authority, and the Panchayats (District Councils).

The Institute of Wetland Management and Ecological Design was set up by the Government of West Bengal, and among other projects, it has taken several wetlands initiatives. First, it has used GIS (geographic information systems) technology to create various maps of the wetlands in order to analyze the state of each zone, and better understand the hydrology, land use, topography, and changes to the system over time.

The Institute has also created the East Kolkata Wetlands Management Committee and various sub-committees which are working on an action plan that brings together all "stakeholders" of the wetlands to share their knowledge and cooperate on their preservation. It is also intended to create better technical knowledge and systems for waste management and farming practices as well as addressing legal issues which would prevent conversion of water bodies, specify property ownership, and amend existing land use policies. The project gets funding from DFID (Department for International Development, the UK government's official overseas aid agency), the ADB (Asian Development Bank), and the central government of India.

A series of "multi-stakeholder workshops" (11 local and 3 major) was held to come up with specific proposals for the wetlands. These took place between 2000 and 2003 through interviews with fish, rice and vegetable farmers, community meetings and household interviews.

What emerged were several insights which helped to create an EKW conservation plan. Water management was within all of the larger issues (silt buildup, pollution, decrease of nutrient content, sluice gate management). Solutions proposed during these sessions varied in focus and scale.

For example, some were technical (desilting of canals), government-related (policies), environmental (pollution control), institutional (creation of new agencies), and social (capacity building and training). The work is still in its early stages, but according to Stuart Bunting, the action plan has had good feedback as it has finally brought together and coordinated groups that were working in chaos and cross-purposes, and set the stage for more clarity and organization in the conservation process.

According to a report by the World Bank and the International Development Research Center in Canada, the reuse of sewage in fish culture, algal and aquatic plant production and energy production have been identified as new, promising technologies that radically change the context of urban sanitation. The EKWs are an opportunity for research and study that may influence the development of low-cost sanitation technology, especially for developing countries.

In "Wastewater-Fed Aquaculture in the Wetlands of Calcutta-an Overview," Ghosh writes:

> *"This system shows the significance of reuse in waste disposal and sanitary engineering which has not been well appreciated. It's been shown that integrated resource recovery and institutionalization of urban solid and liquid waste management can reduce costs to municipalities by 30-90%."*

If the wetlands fall prey to urban expansion, the results will be not only the loss of food and livelihood, but will require the construction of an expensive sewage treatment facility to replace the services it currently offers. This case demonstrates the value of "invisible services" such as waste management, food production and livelihood, and it highlights how nature (sun, bacteria, algae, water hyacinths, fish and various fauna) can "do the work" to provide these services on a relatively large scale and in an urban environment.

Therefore in this context, the EKWs, despite their problems, are an excellent example as one of the world's largest systems of resource recovery.

APPROPRIATE AND EFFECTIVE STRATEGY IN WATER HARVESTING

In the face of overall, ever deteriorating water situation in our country, the joint efforts of the NGO, Government, International Organizations, non Government agencies and the community, have revived several local rivers and rivulets (62 till September 2007) through the series of small scale water harvesting structures in the tribal regions and other parts of western India. The technology and approach of this development have potential for the widespread replication in the entire country in general and in the tribal regions in particular.

River-II is a local small river partly flowing through Dahod Tehsil and partly Jhalod Tehsil of District Dahod, Gujarat. It is a predominantly tribal area. Several tribal villages are situated along the banks of this river. This river and the people around it have witnessed three sharp changes in the last half-century. The elderly people telling us the story of this river being perennial before 50 years. After that till recent times, suddenly the water disappears from this river for about 6-8 months in a year. After monsoon, i.e. from October to December, the river flow would be nominal, and come January and there would be no flow with the entire river drying up.

But, in the recent and latest development, the entire river is almost full of water even during the driest summer months of April-May-June. This is not due to any magic. This amazing change has been brought in by the appropriate technological interventions in the form of series of small masonry water harvesting structures (small check dams) at the appropriate distance across the entire length of the river. People on the two banks of this river have very interesting observations and equally interesting new demands. They tell us that before this series of check dams, they were crossing the river at any point from October onwards to the monsoon season (from October to June). But, now when they reach the river, they have to think from where to cross it. They are infact, very happy with this development seeing their river full of water during the summer, but, they now want culverts/bridges along the river so that they can easily cross the river. They have now placed their demand for such culverts on this river. Meanwhile, improvising the design of the check dams, the implementing NGO has provided cause-way on most of the check dams so that atleast people and cattles can cross the river easily.

The impact of this series of water harvesting structures on the river has been astonishing in the form of increased irrigation, availability of water round the year for people and cattle, increase in subsoil water level in the adjacent area and above all due to the availability of water and food self sufficiency, the seasonal migration rate which was around 70 % of the population of any village on this river has come down to below 10 %. On this river, there are now 18 community lift irrigation systems installed, irrigating 3,100 acres of the land in Rabi alone. This development took place because of series of water harvesting structures in such a small source, offering benefit to all the villages situated on this river.

River Kali-II is not an isolated example of almost optimum water harvesting. Khan, Kali-I, Dudhimati, Machhan, Hadaf, Panam, Walwa, Welwa, Wankdi, Ujol, Chandroi, etc. in Gujarat and Pat, Sangus, Ranek, Negdi, Hiren, Jiradhara, Telni, Pundiya, etc. in Banswara, Rajasthan, Kshipra, Chambal, Kalisindh, Ahu in Jhalawar, Rajasthan, totally about 62 local rivers, rivulets have been almost optimally harvested through the series of structures on these water sources mainly in about 75 kms radius (excepting Jhalawar region), cutting across the borders of 3 States, Gujarat, Rajasthan and Madhya

Pradesh. These 62 local rivers and rivulets have been revived, making them perennial. More such sources (small rivers and rivulets) are being developed under this approach every year, and thus, the process of revival of rivers and rivulets has been going on continuously and effectively under Sadguru approach.

In the latest development, big rivers have been harvested with small water harvesting structures without affecting people or property. Big rivers like Chambal and Kshipra have been recently harvested with small structures. On river Kshipra a border river of Rajasthan, Madhya Pradesh, five such water harvesting structures have been raised and as a result, in the driest of the summer months (March-June 2003) this river was full of water for more than 27 kilometres at a stretch, against the scenario that elsewhere in this river there was no water during above summer months. This approach of small structures on big rivers has a special and significant message to the irrigation planners of our country and elsewhere.

Upstream-Downstream Development

In the water resources development and management, the issue of upstream and downstream development has not been attended seriously by the irrigation planners and the government as a whole. Particularly, when big system is created, it leaves upstream and downstream of the source high and dry. Such big structure creates almost monopoly of the source at the cost of entire upstream and downstream region. All principles of equity and riparian rights are thrown to the winds in not allowing the structures often small structures, in upstream. This policy is utterly unjust, un-called for and technically also not desirable. There are technical solutions to such unjust policy. When it is utmost necessary, the state may go for big system. No problem.

But, at the same time small structures in the upstream need to be permitted to meet the needs of upstream inhabitants. Even if it amounts to somewhat less water going to big system, such small structures have to be offered for the upstream villages which have suffered due to the creation of big system. There could be a practical formula that the entire monsoon flow will go to the big system and post monsoon flow could be harvested by small harvesting structures with gated structures. The deficit in the command of big system could be as well compensated by the area irrigated from such small structures. There is big advantage in such small structures that they would be managed by the farmers themselves or their organizations, unlike big systems. This approach will also revive the river which has dried up because of big structures. Also, when there are series of small structures upstream of big system, the base flow will improve. The use of upstream structures would be invariably by lift irrigation systems, which would be much more efficient than the flow irrigation.

The structures on downstream of big systems should have no problem with the big system upstream. The overflow of big system, the seepages of big system and more importantly the return flow of the canals of big system coming back to the river system, will ensure recharging of small structures downstream of big systems. We all know that in big systems often the canals flow without use at night, and therefore, such wastage of water could be brought back to the river system to benefit the downstream villages not served by the canal system. It may be possible at many places that canals length could be reduced if such small structures can serve part of the command of big system.

The above approach of the development of water resources through small structures on upstream and downstream of big systems would be very helpful in the irrigation management as a whole, serving more needy in more efficient manner.

Sadguru Foundation has adopted above approach on downstream and upstream of big and medium systems of the government and the approach has been proved very successful to the farmers.

Drought and Check Dams

Even during the consecutive four years of drought (1999 to 2002), an unusual phenomenon in hundred year's history of the region, most of the check dams had adequate water supply for the people and cattle dependent on the river. The consecutive drought of four years had its own adverse impact on check dams and water storage. However, due to series of check dams on the same source, the sub-soil water has increased and as a result, large numbers of open dug wells have been recharged. In new and innovative approach, open dug wells are constructed downstream of the check dams, so that even if there is no water in check dam, there is enough water in such open dug wells taking care of the drinking water requirement of the people.

Large number of these water-harvesting structures, though small, have capacity to store water ranging from 10-26 mcft in most of the check dams and in recently constructed check dams on big rivers like Chambal and Kshipra, the storage capacity ranges from 40-90 mcft in each check dam. This is sizeable storage. Another very important aspect of our structures is that comparing the ultimate benefits in the form of storage capacity, their cost is one fifth of the similar structures constructed by the Government. High degree of specialization and acumen acquired by our highly qualified and richly experienced engineers get reflected in such amazing cost effectiveness of our structures. Our structures are both useful and beautiful, reflecting the aesthetic sense of the engineers and organization.

At the time of preparing this document in October 2005, the biggest check dam though small in technical definition is being constructed on river Mahi on the border of Banswara and Dungarpur district at Baneshwardham at the cost of Rs. 4.50 crores which will store water of about 310 mcft, a huge storage by the standard of small structure.

Environmental Aspect

All these water-harvesting structures are harmless to anyone from any aspect and they are fully environment friendly. Not a single human being is uprooted, not an inch of land has been submerged, no damage or loss of property, tree or forest under this approach. The approach has, infact, expanded the natural resource base.

So much is the positive impact of these check dams, along with the good watershed development programme that in our project area the ground water table has increased in recent years as acknowledged by the study by the Central Ground Water Board and other studies. We are told this is the only sub-region in Gujarat, which has witnessed rise in the ground water, quite in contrast to the situation in the entire State where ground water has been alarmingly depleted.

Equity

In these days lot of people talk on equity in any development programme, sometime just out of fashion and sometime genuinely. The author believes that the perfect equity is perhaps utopian phenomenon, which is easier to talk than to achieve. It is an ideal, which may not be possible to achieve in each and every field and at each and every place. Yet, by and large in the water harvesting approach of Sadguru Foundation, fair degree of equity is observed. When the entire river is harvested and each and every village situated along the river has got the benefit of water stored through the series of structures, the concept of equity to the extent achievable is ensured.

Also, the villages where these check dams are constructed are not the part of any medium or major irrigation system. Thus, these check dams meet the needs of people who are deprived of major and medium system.

Another equity aspect of these check dams is that they are all benefiting the poorest of the people (our tribal people) and thereby these check dams not only benefit to the most dis-advantaged and deprived people, but also bridge the imbalance between the progressive regions and backward regions.

Community Management

All these water harvesting structures, though technically small, each benefits at an average more than 100 families. In many a cases, relatively bigger structures, though technically small, serves more than 500 families. Local peoples initiative, participation and management are total in the entire process of such structures. The structures are created at the initiative of the local community, the sites are selected according to peoples' opinion, subject to the technical feasibility. Their involvement in the construction and the post construction management is total. Under the technical guidance of NGO, local non-technical supervisors are trained to accomplish these structures. Two persons selected by the community, one man and one woman, supervise the entire process of construction. The post construction management is handed over to the local community. The structures on which lift irrigation schemes are installed and managed by the lift irrigation co-operatives as the beneficiaries of the check dams and lift irrigation scheme are the same and where there is no lift irrigation scheme, the structures are managed by the community forming a group. The post construction reparation, maintenance and management are fully looked after by the community.

Potential

What has been done in the tribal region of Western India by one NGO in the field of water resources, particularly, in water harvesting is of great significance, but, much greater significance of this development lies in the tremendous potential of similar development in all the tribal regions of our country, particularly, the more vulnerable and extremely poor tribal regions in the Central-West part of our country.

About 75 % of the total tribal population of our country is concentrated in 8 States, namely, Bihar, Jharkhand, Orissa, Chattisgarh, Madhya Pradesh, Gujarat, Rajasthan and Maharashtra, broadly located in the Central-West and forming almost contiguous tribal belt.

By any yardstick or indicators the tribal regions of these 8 States are among the most backward and poor. Yet, paradoxically, all these regions are rich in natural resources, and therefore, these regions and the people should not have been so poor. I am not going to deal with the minerals of these regions. Its a different story by itself, in which the mining and the industrial development based on such minerals have not helped very much to the local tribals, which incidentally is a pointer to those who think that industrial development is the only solution for the removal of tribal poverty.

Talking on the potential for the water harvesting in these 8 States, particularly, in the tribal regions of these States, the potential lies in the average rainfall, which is about 1,000 mm, with minimum being 800 mm and maximum touching 1,600 mm. Combining this rainfall with the topography and network of drainage systems, it offers a great scope for water harvesting through the structures, mainly small structures.

There are 47 Districts in above 8 Central-West States, which have either the predominance of tribal population or sizeable tribal population in part of some Districts. Considering the rainfall, rivers, rivulets and entire drainage systems in these regions, presumably, there could be potential for approximately 45,000 small water harvesting structures. Based on the experience in the region, at an average about 100 families may benefit from one water harvesting structure. Thus, on full exploitation of the potential, about 46,00,000 (4.6 million) families and about 2,50,00,000 (25 million) people could benefit. This single programme of water harvesting on such a large scale could benefit sizeable tribal population of the regions. If the big rivers are tapped with series of Bandharas and the entire programme is integrated with watershed development, emphasizing on tree plantation on the suitable land, majority of the tribal population of the entire region could be immensely benefited. Water and trees put together could benefit more than 5 crores (50 million) of people, mostly tribals, in the region through the environmental friendly programme, enriching the environment and people.

On the potential for the small scale water harvesting structures at the macro level of nation as a whole, adding remaining more than 300 non-tribal districts, there could be scope for about 2,00,000 (0.2 million) small scale water harvesting structures on rivers, rivulets and nallas. Considering 25 % over estimation, then also the potential is enormous.

The question is how to transform this potential into concrete development. Who will do it and how ? It is also imperative that the planners and the Governments acknowledge the potential for the small-scale water harvesting structures and give highest priority in the State and National planning. We cannot survive without water and we just cannot ignore the potential for the small-scale water harvesting structures.

In India for any farmer and community, the irrigation is number one priority. Hence, wherever there is potential for irrigation, it has to be accorded the highest priority. In our enthusiasm for new ideas, new approaches, the basic need of people should not be forgotten by the donors and the planners.

We are at the fringe of peril in the water sector and have already reached an alarming situation. The four successive drought years (1999 to 2002) have exposed our planning and implementation of water resources development. Our cities and villages are just not in a position to withstand the onslaught of water scarcity. The present is dark and the future will create unimaginable and unprecedented situation, unless appropriate and speedy measures are taken in water harvesting and water management. Any technology, traditional or modern, any method of harvesting, relevant to the specific region and capable of bringing in effective and sustainable results, has to be adopted for the future of this country vis-a-vis the water requirements. Any bias for or against any size, system or technology in water harvesting field will not do good to the country, where large parts face recurring and often chronic shortage of water. It is a challenging situation, but not hopeless if we muster our resources and plan and implement sound programmes, based on several successful experiences.

CLIMATE CHANGE AND RAINWATER HARVESTING

Holocene (~ 11 kyr BP to present) climate fluctuations, including large spatial variation in Holocene monsoon and temperature over India are now well-resolved from various climate change proxies. A somewhat coherent, although unresolved, picture that emerges for India is as follows: as the aridity increased in the region as is evident from palaeoclimatic studies, people intensified rainwater harvesting as is seen from archaeological and historical evidences.

Folk sayings such as "capture rain where it falls" may have originated in response to the increased aridity in the Indian region over the last few millennia. Such climate fluctuations may have given rise to traditional village tanks, ponds and earthen embankments numbering more than 1.5 million that still harvest rainwater in 660,000 villages in India and encourage growth of vegetation in commons and agro-ecosystems. (Deep Narayan Pandey, et al., CURRENT SCIENCE, VOL. 85, NO. 1, 10 JULY 2003) Indeed, a specific emphasis through the long sweep of history on management of rainwater harvesting systems in ancient texts, such as *Rigveda* (1500 BC), *Atharva Veda* (800 BC), Kautilya's *Arthasastra* (300 BC), Varahamihira's *Brihatsamhita* (AD 550), and Kalhana's *Rajatarangini* (AD 1148' 1150) may document adaptation to fluctuating climate. There is other evidence of climate change rainwater harvesting hypothesis. Majority of palaces and forts (perhaps all) constructed during the 13-18th century developed elaborate water-harvesting systems.

Addressing water problem holds the promise in future for a world compounded by climate change, growing population and decreasing water-impounding area of traditional tanks due to urbanization. We will have to take into account the large-scale, natural climate variations as well as human-induced climate change in the management of natural, social and economic systems. Alternative to ecologically damaging, socially intrusive and capital-intensive water management projects that fail to deliver their desired benefits, it would be useful investing in decentralized facilities, efficient technologies and policies, and human capital to improve overall productivity rather than to find new sources of water supply. Such efforts would need to be encouraged with innovative policy regimes that concurrently promote rainwater harvesting.

Indian Monsoon

Monsoon is a season looked forward with great anticipation all over India. Cool winds begin to blow from the southwest often with gale-like intensity, followed soon after by welcome showers of copious rain. Skies remain cloudy for nearly three months, the whole scene of India gets transformed and tremendous activity is witnessed all over the country. Sometimes, the starting of the southwesterly winds may get delayed, the quantum of water received may get reduced and there may be long breaks in the monsoon current causing anxiety among the people wholly dependent on this heaven-born resource. Monsoon, therefore, is of vital concern to all of us.

During the summer months (June-September) cool winds charged with moisture blow from the SW. Due to the existence of the high Sahya-dri barrier across their flow path, the low-lying moisture-rich currents are forced to rise, become cooled, condense and precipitate in the form of rain on the windward slopes of the Sahya-dri. The *Sahyadri* has played an important role in fashioning the climate of India and the timing of elevation of this mountain range is therefore crucial to our understanding the periodicity and predictability of the Indian monsoon. Similarly, the winds blowing northward over the Bay of Bengal are obstructed and diverted westwards by the Himalaya causing copious amounts of rainfall all along the foothill region and the Gangetic valley in front of the mountains. These features point to the important role played by oceans, and the major mountain ranges of India the Sahya-dri and the Himalaya in the distribution of rainfall which is characteristically Indian and termed therefore the Indian summer monsoon. (Geological Society of India, Selected Editorials)

Water Harvesting in Ancient India

Traditionally, such systems have been integrated with agro-forestry and ethno-forestry practices, and remain useful in contemporary conservation and ecological restoration of degraded ecosystems. A systematic support to local innovations on rainwater harvesting could provide substantial amounts of water. Rural and urban water use, restoration of streams for recreation, freshwater fisheries, and protection of natural ecosystems are all competing for water resources earlier dedicated to food production. Under such circumstances, decentralized rainwater harvesting adaptations prove efficient.

Traditional systems would become more efficient if scientific attempts are combined to enhance the productivity of local knowledge. With an insightful policy, rainwater harvesting can be promoted as a core adaptation strategy for achieving the global security and sustainability of water resources in an era of anthropogenic climate change. Rainwater harvesting in response to climate extremes enhances the resilience of human society. In a world confronting local and global changes, building resilience of human society to absorb shock, learn and develop, would depend on sound knowledge of the historical adaptive processes that are still functional. An integrated perspective of traditional knowledge on adaptation strategies, such as the rainwater harvesting system, is particularly useful to comprehend vulnerability and adaptation to environmental stresses at the local scale. I believe that neither the water policy nor the climate policy discussions seem to notice the worth of rainwater harvesting as an adaptation to climate change, especially in urban areas where water resources are fast depleting due to rapid increase in population and unrestricted use of water. Studies of historical societal adaptations to climate fluctuations provide insights into possible responses of modern societies to future climate change and sustainable management of water resources

Flooding, a Common Feature of Urban Calendar

The core of the present SWD system in city (Mumbai) is about 70 year old, comprising of about 400 km of underground drains and laterals built on the basis of population and weather conditions. The old SWD system is capable of handling rain intensity of 25 mm per hour at low tide. If the rain intensity is more than 25 mm per hour and high tide occurs, there is always a possibility of water logging.

Since the discharge of all the storm water and treated sewage is into the Arabian Sea, tidal variation has a major bearing in the system of storm water drainage (SWD) resulting in flooding and water logging during heavy rains and recession of water during low tide.

There was heavy rainfall in July 2005 when entire City was flooded rail & road traffic was disrupted and industry suffered heavy losses.

WATER HARVESTING, A REVIEW OF DIFFERENT TECHNIQUES

Water harvesting as a traditional but recently improved strategy to make agricultural production in dry areas more reliable is becoming, especially after the recent drought period in the Sahel, more and more a topic of interest. From different traditional agricultural societies in dry areas very successful examples of water harvesting are known. Still the results of recent water harvesting experiences in different development projects are not always positive. ILEIA tried to find which constraints and possibilities water harvesting really has for nowadays agriculture in dry areas.

Coen Reijntjes

Definition and Benefits

"Water harvesting" is the general name used for all the different techniques to collect runoff or flood water for storage in the soil profile or in tanks so that it can be used for the production of crops, trees or fodder. "Water harvesting" also can be the collection of runoff water for human or livestock consumption. In this article this type of water harvesting will not be discussed.

The benefit of water harvesting is not only to secure and increase crop production in semi-arid regions where rainfall is normally high enough for crop production or to make crop production possible in regions were rainfall is normally not sufficient, but also to stop soil erosion and to re charge aquifers tapped for irrigation. An underestimated benefit of water harvesting is also the improvement of soil fertility. Silt, manure and other organic matter is "harvested" or kept in place together with the water. The soil profile stays moist for a longer time, which stimulates soil life so that the formation of stable humus, the nutrient availability, and the water holding capacity are improved. A classic example of water and nutrient harvesting was the delta of the river Nile before the Aswan dam was constructed. The yearly flooding of agricultural fields made permanent agriculture in this arid region possible.

After the construction of the Aswan dam, this was no longer possible. Now farmers have, besides advantages of greater water availability, big problems in keeping the land fertile. Small water harvesting systems with an external catchment area are functioning more or less in the same way. Water harvesting systems with a "within-field" catchment are, of course, harvesting less nutrients but have the advantage of erosion control in situ.

Classification

The different water harvesting systems can be classified in the following way:

A – Systems with an external catchment area for collection of runoff water or flood water from small watersheds:

A1 – agricultural use, without any special arrangements, of natural depressions where runoff or flood water is concentrated temporary and water infiltration is relative high (traditional in e.g. West and East-Africa);

A2 – simple techniques for water spreading and infiltration by means of low, permeable bunds (ridges) which follow the contour lines. These bunds are made of stones, bundled sticks, crop residues or fences of living plants;

A3 – "water pockets"-holes for seeding, runoff collection and management of organic matter ("zai" in Burkina Faso, "covas" in Cape Verde, both traditional);

A4 – half circular or "V" shaped ridges mainly used for tree planting and rangeland improvement, new;

A5 – "Water collection", graded bunds or furrows are used for diverting runoff from agricultural fields, village and waste lands to tanks situated at a lower level. This water is used for supplementary irrigation in dry periods or as full irrigation (traditional in India, new in e.g. West Africa);

A6 – "Runoff farming", runoff water from a treated (e.g. by a chemical spray or the clearing of gravel-stones to improve the formation of a surface crust to increase the runoff) or untreated catchment area is diverted to lower situated agricultural fields;

A7-"Runoff farming", runoff water and silt from small watersheds is captured by dams in seasonal streambeds or is diverted to agricultural fields. In front of these dams the silt builds terraces which are used for agriculture. The infiltrated water makes crop production possible.

B – Systems for storage and agricultural use of flood waters;

B1 – "Flood water farming": these traditional systems make use of the runoff concentrated by natural water sheds in seasonal or permanent river systems. The flood water is diverted from its natural channel by dams or barrages and led to the agricultural fields were the water is kept impounded by earthen dams around the fields. The infiltrated water is used for agriculture (traditional in e.g. the Nile Delta before the Aswan dam, North India, "ahars and khadirs", South Pakistan, "sailabas or kurkabas"). These systems, which are rather complex and about which we have no reliable data available, will not be discussed in the next articles.

C – Systems with a-within-field-catchment area called-in-situ-water harvesting or-micro catchments-:

C1 – "Negarim", runoff from a small plot (micro-or within catchment) is captured at one side where it infiltrates the soil and directly contributes to the available moisture in the rooted profile of an individual productive tree or shrub;

C2 – "contour ridges or bunds", the same system as C1 but instead of small plots, strips are used. Crops can be seeded in front of the bunds were water infiltration is concentrated;

C3 – "contour beds", the same system as C2, but the beds are "W"-shaped, with alternating wide and narrow ridges. The wide ridges serve as catchment zone, the narrow ridges as planting zone and the furrows can serve as drainage or irrigation channels/mechanisation can be used.

The next three articles discuss experiences with water harvesting in Tunisia, Kenya and India, and the constraints and possibilities of water harvesting for small farmers in the semi-arid tropical zone.

RAIN WATER TANKS-LIFE CYCLE ANALYSIS

A typical Australian household uses about 280,000 litres of water annually. Approximately half of this amount is used to water the garden and flush toilets. Installing a rain water tank has the potential to reduce residential fresh water requirements by 50% and contribute to more effective use of precious water resources. In addition, rain water tanks contribute to a reduction in peak stormwater flows and sediment transport. At present, there are three generic types of rain water tanks: reinforced concrete, plastic and galvanised steel tanks.

An independent study, conducted by the Centre for Sustainable Technology at the University of Newcastle, presents a comparison based on Life Cycle Analysis (LCA) of the three types of rain water tanks. The LCA includes all stages in the life of the tank: extraction and processing of raw materials, transport to tank manufacturing site, tank fabrication, transport to customer's site, installation, operation and disposal at the end of the useful life.

5

Ground Water Pollution

Ground-water is water located beneath the ground surface in soil pore spaces and in the fractures of rock formations. A unit of rock or an unconsolidated deposit is called an aquifer when it can yield a usable quantity of water. The depth at which soil pore spaces or fractures and voids in rock become completely saturated with water is called the water table. Ground-water is recharged from, and eventually flows to, the surface naturally; natural discharge often occurs at springs and seeps, and can form oases or wetlands. Ground-water is also often withdrawn for agricultural, municipal and industrial use by constructing and operating extraction wells. The study of the distribution and movement of ground-water is hydrogeology, also called ground-water hydrology.

Typically, ground-water is thought of as liquid water flowing through shallow aquifers, but technically it can also include soil moisture, permafrost (frozen soil), immobile water in very low permeability bedrock, and deep geothermal or oil formation water. Ground-water is hypothesized to provide lubrication that can possibly influence the movement of faults. It is likely that much of the Earth's subsurface contains some water, which may be mixed with other fluids in some instances. Ground-water may not be confined only to the Earth. The formation of some of the landforms observed on Mars may have been influenced by ground-water. There is also evidence that liquid water may also exist in the subsurface of Jupiter's moon Europa.

Water pollution is any physical, chemical or biological alteration of water that is harmful to living organisms. A more human-centred definition describes pollution as "any contamination of water that makes it unsuitable for its 'highest' use". For example, silt or mud in water would not affect its use for boating or fishing, but the same silt in the same water could prevent humans from drinking or washing clothes in it. There are many types of water pollution; some have a natural origin, but most are related to human activities.

Water becomes polluted easily because it is a nearly "universal" solvent. One of the alchemists' goals was to find a "universal solvent"-a liquid into which everything could be dissolved. They were unaware that they already possessed what they were looking for-almost everything will dissolve in water, at least to some extent.

This makes water an excellent cleaner, and a convenient way to dispose of waste. These initially positive uses, in turn, often lead to water pollution. There are many sources of water pollution, but a common way to categorize them is to consider point sources and non-point sources. Point sources are specific locations that are easy to identify. Less specific locations, more generalized and harder to find, are called non-point sources, and are often difficult to define and to regulate. Whatever its cause, water pollution does have many possible solutions. Pollution in streams and lakes is often easy to see or smell, and frequently the solutions are obvious. Groundwater, though hidden below the surface, can also become polluted; and because it is harder to see and easier to forget, this pollution is more difficult to detect and to clean up.

Water pollution is a large set of adverse effects upon water bodies (lakes, rivers, oceans, groundwater) caused by human activities. Although natural phenomena such as volcanoes, storms, earthquakes etc. also cause major changes in water quality and the ecological status of water, these are not deemed to be pollution. Water pollution has many causes and characteristics. Increases in nutrient loading may lead to eutrophication.

Organic wastes such as sewage and farm waste impose high oxygen demands on the receiving water leading to oxygen depletion with potentially severe impacts on the whole eco-system. Industries discharge a variety of pollutants in their wastewater including heavy metals, organic toxins, oils, nutrients, and solids. Discharges can also have thermal effects, especially those from power stations, and these too reduce the available oxygen.

Silt-bearing runoff from many activities including construction sites, deforestation and agriculture can inhibit the penetration of sunlight through the water column, restricting photosynthesis and causing blanketing of the lake or river bed, in turn damaging ecological systems. Pollutants in water include a wide spectrum of chemicals, pathogens, and physical chemistry or sensory changes, though the main cause of pollution in both lakes and rivers are animal wastes and fertilizer. Many of the chemical substances are toxic or even carcinogenic. Pathogens can obviously produce waterborne diseases in either human or animal hosts. Alteration of water's physical chemistry include acidity, conductivity, temperature, and eutrophication. Eutrophication is the fertilisation of surface water by nutrients that were previously scarce. Even many of the municipal water supplies in developed countries can present health risks. Water pollution is a serious problem in the global context. It has been suggested that it is the leading worldwide cause of deaths and diseases, and that it accounts for the deaths of more than 14,000 people daily.

Sources of Water Pollution

Principal sources of water pollution are:

- industrial discharge of chemical wastes and byproducts
- discharge of poorly-treated or untreated sewage
- surface runoff containing pesticides
- slash and burn farming practice, which is often an element within shifting cultivation agricultural systems
- surface runoff containing spilled petroleum products
- surface runoff from construction sites, farms, or paved and other impervious surfaces e.g. silt
- discharge of contaminated and/or heated water used for industrial processes
- acid rain caused by industrial discharge of sulphur dioxide (by burning high-sulphur fossil fuels)
- excess nutrients added by runoff containing detergents or fertilizers
- underground storage tank leakage, leading to soil contamination, thence aquifer contamination.

Contaminants

Contaminants may include organic and inorganic substances.

Some organic water pollutants are:

- insecticides and herbicides, a huge range of organohalide and other chemicals
- bacteria, often is from sewage or livestock operations;
- food processing waste, including pathogens
- tree and brush debris from logging operations
- VOCs (Volatile Organic Compounds, industrial solvents) from improper storage

Some inorganic water pollutants include:

- heavy metals including acid mine drainage
- acidity caused by industrial discharges (especially sulphur dioxide from power plants)
- chemical waste as industrial by products
- fertilizers, in runoff from agriculture including nitrates and phosphates

- silt in surface runoff from construction sites, logging, slash and burn practices or land clearing sites

Transport and Chemical Reactions of Water Pollutants

Most water pollutants are eventually carried by the rivers into the oceans. In some areas of the world the influence can be traced hundred miles from the mouth by studies using hydrology transport models. Advanced computer models such as SWMM or the DSSAM Model have been used in many locations worldwide to examine the fate of pollutants in aquatic systems. Indicator filter feeding species such as copepods have also been used to study pollutant fates in the New York Bight, for example. The highest toxin loads are not directly at the mouth of the Hudson River, but 100 kilometres south, since several days are required for incorporation into planktonic tissue. The Hudson discharge flows south along the coast due to coriolis force. Further south then are areas of oxygen depletion, caused by chemicals using up oxygen and by algae blooms, caused by excess nutrients from algal cell death and decomposition. Fish and shellfish kills have been reported, because toxins climb the foodchain after small fish consume copepods, then large fish eat smaller fish, etc. Each step up the food chain concentrates certain toxins like heavy metals and DDT by approximately a factor of ten.

For several years ocean researcher Charles Moore has been investigating a concentration of floating plastic debris in the Pacific Ocean. His study indicates that ocean currents have added to the mass until it is now about the size of Texas. Many of these long-lasting pieces wind up in the stomachs of marine birds and animals.

Many chemicals undergo reactive decay or change especially over long periods of time in groundwater reservoirs. A noteworthy class of such chemicals are the chlorinated hydrocarbons such as trichloroethylene (used in industrial metal degreasing) and tetrachloroethylene used in the dry cleaning industry. Both of these chemicals, which are carcinogens themselves, undergo partial decomposition reactions leading to new hazardous chemicals.

Groundwater pollution is much more difficult to abate than surface pollution because groundwater can move great distances through unseen aquifers. Non-porous aquifers such as clays partially purify water of bacteria by simple filtration (adsorption and absorption), dilution, and, in some cases, chemical reactions and biological activity: however, in some cases, the pollutants merely transform to soil contaminants. Groundwater that moves through cracks and caverns is not filtered and can be transported as easily as surface water. In fact this can be aggravated by the human tendency to use natural sinkholes as dumps in areas of Karst topography.

There are a variety of secondary effects stemming not from the original pollutant, but a derivative condition. Some of these secondary impacts are:

- Silt bearing surface runoff from can inhibit the penetration of sunlight through the water column, hampering Photosynthesis in aquatic plants.
- Thermal pollution can induce fish kills and invasion by new thermophyllic species

Regulatory Framework

In the UK there are common law rights to protect the passage of water across land unfettered in either quality of quantity. Criminal laws dating back to the 16th century excercised some control over water pollution but it was not until the River (Prevention of pollution) Acts 1951-1961 were enacted that any systematic control over water pollution was established. These laws were strengthened and extended in the Control of Pollution Act 1984 which has since been updated and modified by a series of further acts. In the USA, concern over water pollution resulted in the enactment of state anti-pollution laws in the latter half of the 19th century, and federal legislation enacted in 1899. The Refuse Act of the federal Rivers and Harbors Act of 1899 prohibits the disposal of any refuse matter from into either the nation's navigable rivers, lakes, streams, and other navigable bodies of water, or any tributary to such waters, unless one has first obtained a permit. The Water Pollution Control Act, passed in 1948, gave authority to the Surgeon General to reduce water pollution.

Growing public awareness and concern for controlling water pollution led to enactment of the Federal Water Pollution Control Act Amendments of 1972. As amended in 1977, this law became commonly known as the Clean Water Act. The Act established the basic mechanisms for regulating contaminant discharge. It established the authority for EPA to implement wastewater standards for industry. The Clean Water Act also continued requirements to set water quality standards for all contaminants in surface waters. Further amplification of the Act continued including the enactment of the Great Lakes Legacy Act of 2002.

AQUIFERS

An *aquifer* is a layer of porous substrate that contains and transmits ground-water. When water can flow directly between the surface and the saturated zone of an aquifer, the aquifer is unconfined. The deeper parts of unconfined aquifers are usually more saturated since gravity causes water to flow downward. The upper level of this saturated layer of an unconfined aquifer is called the *water table* or *phreatic surface*. Below the water table, where generally all pore spaces are saturated with water is the phreatic zone. Substrate with low porosity that permits limited transmission of ground-water is known as an *aquitard*. An *aquiclude* is a substrate with porosity that is so low it is virtually impermeable to ground-water.

A *confined aquifer* is an aquifer that is overlain by a relatively impermeable layer of rock or substrate such as an aquiclude or aquitard. If a confined aquifer follows a downward grade from its *recharge zone*, ground-water can become pressurized as it flows. This can create artesian wells that flow freely without the need of a pump and rise to a higher elevation than the static water table at the above, unconfined, aquifer.

The characteristics of aquifers vary with the geology and structure of the substrate and topography in which they occur. Generally, the more productive aquifers occur in sedimentary geologic formations. By comparison, weathered and fractured crystalline rocks yield smaller quantities of ground-water in many environments. Unconsolidated to poorly cemented alluvial materials that have accumulated as valley-filling sediments in major river valleys and geologically subsiding structural basins are included among the most productive sources of ground-water.

The high specific heat capacity of water and the insulating effect of soil and rock can mitigate the effects of climate and maintain ground-water at a relatively steady temperature. In some places where ground-water temperatures are maintained by this effect at about 50°F/10°C, ground-water can be used for controlling the temperature inside structures at the surface.

For example, during hot weather relatively cool ground-water can be pumped through radiators in a home and then returned to the ground in another well. During cold seasons, because it is relatively warm, the water can be used in the same way as a source of heat for heat pumps that is much more efficient than using air.

Water Cycle

Ground-water makes up about twenty percent of the world's fresh water supply, which is about 0.61% of the entire world's water, including oceans and permanent ice. Global ground-water storage is roughly equal to the total amount of freshwater stored in the snow and ice pack, including the north and south poles. This makes it an important resource which can act as a natural storage that can buffer against shortages of surface water, as in during times of drought.

Ground-water is naturally replenished by surface water from precipitation, streams, and rivers when this recharge reaches the water table. Ground-water can be a long-term 'reservoir' of the natural water cycle (with residence times from days to millennia), as opposed to short-term water reservoirs like the atmosphere and fresh surface water (which have residence times from minutes to years).

The Great Artesian Basin in central and eastern Australia is one of the largest confined aquifer systems in the world, extending for almost 2 million km^2. By analysing the trace elements in water sourced from deep underground, hydrogeologists have been able to determine that water extracted from these aquifers can be more than 1 million years old.

By comparing the age of ground-water obtained from different parts of the Great Artesian Basin, hydrogeologists have found it increases in age across the basin. Where water recharges the aquifers along the Eastern Divide, ages are young. As ground-water flows westward across the continent, it increases in age, with the oldest ground-water occurring in the western parts. This means that in order to have travelled almost 1000 km from the source of recharge in 1 million years, the ground-water flowing through the Great Artesian Basin travels at an average rate of about 1 metre per year.

Issues

Overview

Certain problems have beset the use of ground-water around the world. Just as river waters have been over-used and polluted in many parts of the world, so too have aquifers. The big difference is that aquifers are out of sight. The other major problem is that water management agencies, when calculating the 'sustainable yield' of aquifer and river water, have often counted the same water twice, once in the aquifer, and once in its connected river. This problem, although understood for centuries, has persisted, partly through inertia within government agencies. In Australia, for example, prior to the statutory reforms initiated by the Council of Australian Governments water reform framework in the 1990s, many Australian States managed ground-water and surface water through separate government agencies, an approach beset by rivalry and poor communication.

The time lags inherent in the dynamic response of ground-water to development have generally been ignored by water management agencies, decades after scientific understanding of the issue was consolidated. In brief, the effects of ground-water overdraft (although undeniably real) may take decades or centuries to manifest themselves.

In a classic study in 1982, Bredehoeft and colleagues modelled a situation where ground-water extraction in an intermontane basin withdrew the entire annual recharge, leaving 'nothing' for the natural ground-water-dependent vegetation community. Even when the borefield was situated close to the vegetation, 30% of the original vegetation demand could still be met by the lag inherent in the system after 100 years. By year 500 this had reduced to 0%, signalling complete death of the ground-water-dependent vegetation.

The science has been available to make these calculations for decades; however water management agencies have generally ignored effects which will appear outside the rough timeframe of political elections (3 to 5 years). Marios Sophocleous argued strongly that management agencies must define and use appropriate timeframes in ground-water planning. This will mean calculating ground-water withdrawal permits based on predicted effects decades, sometimes centuries in the future.

As water moves through the landscape it collects soluble salts, mainly sodium chloride. Where such water enters the atmosphere through evapotranspiration, these salts are left behind. In irrigation districts, poor drainage of soils and surface aquifers can result in water tables coming to the surface in low-lying areas. Major land degradation problems of soil salinity and waterlogging result, combined with increasing levels of salt in surface waters. As a consequence, major damage has occurred to local economies and environments.

Four important effects are worthy of brief mention. First, flood mitigation schemes, intended to protect infrastructure built on floodplains, have had the unintended consequence of reducing aquifer recharge associated with natural flooding. Second, prolonged depletion of ground-water in extensive aquifers can result in land subsidence, with associated infrastructure damage – as well as (thirdly) saline intrusion. Fourth, draining acid sulphate soils, often found in low-lying coastal plains, can result in acidification and pollution of formerly freshwater and estuarine streams. Another cause for concern is that ground-water drawdown from over-allocated aquifers has the potential to cause severe damage to both terrestrial and aquatic ecosystems – in some cases very conspicuously but in others quite imperceptibly because of the extended period over which the damage occurs.

Overdraft

Ground-water is a highly useful and often abundant resource. However, over-use, or overdraft, can cause major problems to human users and to the environment. The most evident problem (as far as human ground-water use is concerned) is a lowering of the water table beyond the reach of existing wells. Wells must consequently be deepened to reach the ground-water; in some places (e.g., California, Texas and India) the water table has dropped hundreds of feet because of extensive well pumping. In the Punjab region of India, for example, ground-water levels have dropped 10 meters since 1979, and the rate of depletion is accelerating. A lowered water table may, in turn, cause other problems such as ground-water-related subsidence and saltwater intrusion.

Ground-water is also ecologically important. The importance of ground-water to ecosystems is often overlooked, even by freshwater biologists and ecologists. Groundwaters sustain rivers, wetlands and lakes, as well as subterranean ecosystems within karst or alluvial aquifers. Not all ecosystems need ground-water, of course. Some terrestrial ecosystems - for example, those of the open deserts and similar arid environments - exist on irregular rainfall and the moisture it delivers to the soil, supplemented by moisture in the air. While there are other terrestrial ecosystems in more hospitable environments where ground-water plays no central role, ground-water is in fact fundamental to many of the world's major ecosystems. Water flows between groundwaters and surface waters. Most rivers, lakes and wetlands are fed by, and (at other places or times) feed ground-water, to varying degrees. Ground-water feeds soil moisture through percolation, and many terrestrial vegetation communities depend directly on either ground-water or the percolated soil moisture above the aquifer for at least part of each year. Hyporheic zones (the mixing zone of streamwater and ground-water) and riparian zones are examples of ecotones largely or totally dependent on ground-water.

Aquifer drawdown or overdrafting and the pumping of fossil water increases the total amount of water within the hydrosphere subject to transpiration and evaporation processes, thereby causing accretion in water vapour and cloud cover, the primary absorbers of infrared radiation in the Earth's atmosphere. Adding water to the system has a forcing effect on the whole earth system, an accurate estimate of which hydrogeological fact is yet to be quantified.

Subsidence

Subsidence occurs when too much water is pumped out from underground, deflating the space below the above-surface, and thus causing the ground to actually collapse. The result can look like craters on plots of land. This occurs because in its natural equilibrium state, the hydraulic pressure of ground-water in the pore spaces of the aquifer and the aquitard supports some of the weight of the overlying sediments. When ground-water is removed from aquifers by excessive pumping, pore pressures in the aquifer drop and compression of the aquifer may occur. This compression may be partially recoverable if pressures rebound, but much of it is not. When the aquifer gets compressed it may cause land subsidence, a drop in the ground surface. The city of New Orleans, Louisiana, is actually below sea level today, and its subsidence is partly caused by removal of ground-water from the various aquifer/aquitard systems beneath it. In the first half of the 20th century, the city of San Jose, California, dropped 13 feet from land subsidence caused by overpumping; this subsidence has been halted with improved ground-water management.

Seawater Intrusion

Generally, in very humid or undeveloped regions, the shape of the water table mimics the slope of the surface. The recharge zone of an aquifer near the seacoast is likely to be inland, often at considerable distance. In these coastal areas, a lowered water table may induce sea water to reverse the flow toward the land. Sea water moving inland is called a saltwater intrusion. Alternatively, salt from mineral beds may leach into the ground-water of its own accord.

Mining

Sometimes the water movement from the recharge zone to the place where it is withdrawn may take centuries. When the usage of water is greater than the recharge, it is referred to as *mining* water (the water is often called fossil water because of its geologic age). Under those circumstances it is not a renewable resource.

WATER POLLUTION

Water pollution is the contamination of water bodies (e.g. lakes, rivers, oceans and ground-water). Water pollution occurs when pollutants are discharged directly or indirectly into water bodies without adequate treatment to remove harmful compounds.

Water pollution affects plants and organisms living in these bodies of water; and, in almost all cases the effect is damaging not only to individual species and populations, but also to the natural biological communities.

Water pollution is a major global problem. It has been suggested that it is the leading worldwide cause of deaths and diseases, and that it accounts for the deaths of more than 14,000 people daily. An estimated 700 million Indians have no access to a proper toilet, and 1,000 Indian children die of diarrheal sickness every day. Some 90% of China's cities suffer from some degree of water pollution, and nearly 500 million people lack access to safe drinking water. In addition to the acute problems of water pollution in developing countries, industrialized countries continue to struggle with pollution problems as well. In the most recent national report on water quality in the United States, 45 percent of assessed stream miles, 47 percent of assessed lake acres, and 32 percent of assessed bay and estuarine square miles were classified as polluted.

Water is typically referred to as polluted when it is impaired by anthropogenic contaminants and either does not support a human use, such as drinking water, and/or undergoes a marked shift in its ability to support its constituent biotic communities, such as fish. Natural phenomena such as volcanoes, algae blooms, storms, and earthquakes also cause major changes in water quality and the ecological status of water.

Categories

Surface water and ground-water have often been studied and managed as separate resources, although they are interrelated. Surface water seeps through the soil and becomes ground-water. Conversely, ground-water can also feed surface water sources. Sources of surface water pollution are generally grouped into two categories based on their origin.

Point Sources

Point source pollution refers to contaminants that enter a waterway from a single, identifiable source, such as a pipe or ditch. Examples of sources in this category include discharges from a sewage treatment plant, a factory, or a city storm drain. The U.S. Clean Water Act (CWA) defines point source for regulatory enforcement purposes. The CWA definition of point source was amended in 1987 to include municipal storm sewer systems, as well as industrial stormwater, such as from construction sites.

Non–point Sources

Non–point source (NPS) pollution refers to diffuse contamination that does not originate from a single discrete source. NPS pollution is often the cumulative effect of small amounts of contaminants gathered from a large area. A common example is the leaching out of nitrogen compounds from fertilized agricultural lands. Nutrient runoff in stormwater from "sheet flow" over an agricultural field or a forest are also cited as examples of NPS pollution.

Contaminated storm water washed off of parking lots, roads and highways, called urban runoff, is sometimes included under the category of NPS pollution. However, this runoff is typically channeled into storm drain systems and discharged through pipes to local surface waters, and is a point source. However where such water is not channeled and drains directly to ground it is a non-point source.

Ground-water Pollution

Interactions between ground-water and surface water are complex. Consequently, ground-water pollution, sometimes referred to as ground-water contamination, is not as easily classified as surface water pollution. By its very nature, ground-water aquifers are susceptible to contamination from sources that may not directly affect surface water bodies, and the distinction of point vs. non-point source may be irrelevant.

A spill or ongoing releases of chemical or radionuclide contaminants into soil (located away from a surface water body) may not create point source or non-point source pollution, but can contaminate the aquifer below, defined as a toxin plume. The movement of the plume, called a plume front, may be analysed through a hydrological transport model or ground-water model. Analysis of ground-water contamination may focus on the soil characteristics and site geology, hydrogeology, hydrology, and the nature of the contaminants.

HYDROGEOLOGY

Hydrogeology (*hydro-* meaning water, and *-geology* meaning the study of the Earth) is the area of geology that deals with the distribution and movement of ground-water in the soil and rocks of the Earth's crust, (commonly in aquifers). The term geohydrology is often used interchangeably. Some make the minor distinction between a hydrologist or engineer applying themselves to geology (geohydrology), and a geologist applying themselves to hydrology (hydrogeology).

Hydrogeology is an interdisciplinary subject; it can be difficult to account fully for the chemical, physical, biological and even legal interactions between soil, water, nature and society. The study of the interaction between ground-water movement and geology can be quite complex. Ground-water does not always flow in the subsurface down-hill following the surface topography; ground-water follows pressure gradients (flow from high pressure to low) often following fractures and conduits in circuitous paths. Taking into account the interplay of the different facets of a multi-component system often requires knowledge in several diverse fields at both the experimental and theoretical levels. The following is a more traditional introduction to the methods and nomenclature of saturated subsurface hydrology, or simply hydrogeology.

Hydrogeology in Relation to other Fields

Hydrogeology, as stated above, is a branch of the earth sciences dealing with the flow of water through aquifers and other shallow porous media (typically less than 450 m or 1,500 ft below the land surface.) The very shallow flow of water in the subsurface (the upper 3 m or 10 ft) is pertinent to the fields of soil science, agriculture and civil engineering, as well as to hydrogeology.

The general flow of fluids (water, hydrocarbons, geothermal fluids, etc.) in deeper formations is also a concern of geologists, geophysicists and petroleum geologists. Ground-water is a slow-moving, viscous fluid (with a Reynolds number less than unity); many of the empirically derived laws of ground-water flow can be alternately derived in fluid mechanics from the special case of Stokes flow (viscosity and pressure terms, but no inertial term).

The mathematical relationships used to describe the flow of water through porous media are the diffusion and Laplace equations, which have applications in many diverse fields. Steady ground-water flow (Laplace equation) has been simulated using electrical, elastic and heat conduction analogies. Transient ground-water flow is analogous to the diffusion of heat in a solid, therefore some solutions to hydrological problems have been adapted from heat transfer literature. Traditionally, the movement of ground-water has been studied separately from surface water, climatology, and even the chemical and microbiological aspects of hydrogeology (the processes are uncoupled). As the field of hydrogeology matures, the strong interactions between ground-water, surface water, water chemistry, soil moisture and even climate are becoming more clear.

For example: Aquifer drawdown or overdrafting and the pumping of fossil water increases the total amount of water within the hydrosphere subject to transpiration and evaporation processes, thereby causing accretion in water vapour and cloud cover, the primary absorbers of infrared radiation in the earth's atmosphere. Adding water to the system has a forcing effect on the whole earth system. An accurate estimate of the climatic forcing effect due to this hydrogeological fact is yet to be quantified.

Definitions and Material Properties

One of the main tasks a hydrogeologist typically performs is the prediction of future behaviour of an aquifer system, based on analysis of past and present observations. Some hypothetical, but characteristic questions asked would be:

- Can the aquifer support another subdivision?
- Will the river dry up if the farmer doubles his irrigation?
- Did the chemicals from the dry cleaning facility travel through the aquifer to my well and make me sick?

- Will the plume of effluent leaving my neighbor's septic system flow to my drinking water well?

Most of these questions can be addressed through simulation of the hydrologic system (using numerical models or analytic equations). Accurate simulation of the aquifer system requires knowledge of the aquifer properties and boundary conditions. Therefore a common task of the hydrogeologist is determining aquifer properties using aquifer tests.

In order to further characterize aquifers and aquitards some primary and derived physical properties are introduced below. Aquifers are broadly classified as being either confined or unconfined (water table aquifers), and either saturated or unsaturated; the type of aquifer affects what properties control the flow of water in that medium (e.g., the release of water from storage for confined aquifers is related to the storativity, while it is related to the specific yield for unconfined aquifers).

Hydraulic Head

Changes in hydraulic head (h) are the driving force which causes water to move from one place to another. It is composed of pressure head ($\o$) and elevation head (z). The head gradient is the change in hydraulic head per length of flowpath, and appears in Darcy's law as being proportional to the discharge.

Hydraulic head is a directly measurable property that can take on any value (because of the arbitrary datum involved in the z term); ϕ can be measured with a pressure transducer (this value can be negative, e.g., suction, but is positive in saturated aquifers), and z can be measured relative to a surveyed datum (typically the top of the well casing). Commonly, in wells tapping unconfined aquifers the water level in a well is used as a proxy for hydraulic head, assuming there is no vertical gradient of pressure. Often only *changes* in hydraulic head through time are needed, so the constant elevation head term can be left out.

A record of hydraulic head through time at a well is a hydrograph or, the changes in hydraulic head recorded during the pumping of a well in a test are called drawdown.

Porosity

Porosity (n) is a directly measurable aquifer property; it is a fraction between 0 and 1 indicating the amount of pore space between unconsolidated soil particles or within a fractured rock. Typically, the majority of groundwater (and anything dissolved in it) moves through the porosity available to flow (sometimes called effective porosity). Permeability is an expression of the connectedness of the pores. For instance, an unfractured rock unit may have a high *porosity* (it has lots of *holes* between its constituent grains), but a low *permeability* (none of the pores are connected). An example of this phenomenon is pumice, which, when in its unfractured state, can make a poor aquifer.

Porosity does not directly affect the distribution of hydraulic head in an aquifer, but it has a very strong effect on the migration of dissolved contaminants, since it affects ground-water flow velocities through an inversely proportional relationship.

Water Content

Water content is also a directly measurable property; it is the fraction of the total rock which is filled with liquid water. This is also a fraction between 0 and 1, but it must also be less than or equal to the total porosity. The water content is very important in vadose zone hydrology, where the hydraulic conductivity is a strongly nonlinear function of water content; this complicates the solution of the unsaturated ground-water flow equation.

Hydraulic Conductivity

Hydraulic conductivity (K) and transmissivity (T) are indirect aquifer properties (they cannot be measured directly). T is the K integrated over the vertical thickness (b) of the aquifer ($T=Kb$ when K is constant over the entire thickness).

These properties are measures of an aquifer's ability to transmit water. Intrinsic permeability is a secondary medium property which does not depend on the viscosity and density of the fluid (K and T are specific to water); it is used more in the petroleum industry.

Specific Storage and Specific Yield

Specific storage (S_s) and its depth-integrated equivalent, storativity ($S=S_sb$), are indirect aquifer properties (they cannot be measured directly); they indicate the amount of ground-water released from storage due to a unit depressurization of a confined aquifer. They are fractions between 0 and 1.

Specific yield (S_y) is also a ratio between 0 and 1 (S_y d" porosity) and indicates the amount of water released due to drainage from lowering the water table in an unconfined aquifer.

The value for specific yield is less than the value for porosity because some water will remain in the medium even after drainage due to molecular forces. Often the porosity or effective porosity is used as an upper bound to the specific yield. Typically S_y is orders of magnitude larger than S_s.

Contaminant Transport Properties

Often we are interested in how the moving ground-water water will move dissolved contaminants around (the sub-field of contaminant hydrogeology).

The contaminants can be man-made (e.g., petroleum products, nitrate or Chromium) or naturally occurring (e.g., arsenic, salinity). Besides needing to understand where the ground-water is flowing, based on the other hydrologic properties discussed above, there are additional aquifer properties which affect how dissolved contaminants move with ground-water.

Hydrodynamic Dispersion

Hydrodynamic dispersivity is an empirical factor which quantifies how much contaminants stray away from the path of the ground-water which is carrying it. Some of the contaminants will be "behind" or "ahead" the mean ground-water, giving rise to a longitudinal dispersivity, and some will be "to the sides of" the pure advective ground-water flow, leading to a transverse dispersivity.

Dispersion in ground-water is due to the fact that each water "particle", passing beyond a soil particle, must choose where to go, whether left or right or up or down, so that the water "particles" (and their solute) are gradually spread in all directions around the mean path.

This is the "microscopic" mechanism, on the scale of soil particles. More important, on long distances, can be the macroscopic inhomogeneities of the aquifer, which can have regions of larger or smaller permeability, so that some water can find a preferential path in one direction, some other in a different direction, so that the contaminant can be spread in a completely irregular way, like in a (three-dimensional) delta of a river.

Dispersivity is actually a factor which represents our *lack of information* about the system we are simulating. There are many small details about the aquifer which are being averaged when using a macroscopic approach (e.g., tiny beds of gravel and clay in sand aquifers), they manifest themselves as an *apparent* dispersivity. Because of this, α is often claimed to be dependent on the length scale of the problem — the dispersivity found for transport through 1 m^3 of aquifer is different than that for transport through 1 cm^3 of the same aquifer material.

Molecular Diffusion

Diffusion is a fundamental physical phenomenon, by which Einstein explained Brownian motion, which describes the random thermal movement of molecules and small particles in gases and liquids. It is an important phenomenon for small distances (it is essential for the achievement of thermodinamic equilibria), but, as the time necessary to cover a distance by diffusion is proportional to the square of the distance itself, it is ineffective for spreading a solute over macroscopic distances. The diffusion coefficient, D, is typically quite small, and its effect can often be considered negligible (unless ground-water flow velocities are extremely low, as they are in clay aquitards).

It is important not to confuse diffusion with dispersion, as the former is a physical phenomenon and the latter is an empirical factor which is cast into a similar form as diffusion, because we already know how to solve that problem.

Retardation by Adsorption

The retardation factor is another very important feature that make the motion of the contaminant to deviate from the average ground-water motion. It is analogous to the retardation factor of chromatography.

Unlike diffusion and dispersion, which simply spread the contaminant, the retardation factor changes its *global average velocity*, so that it can be much slower than that of water. This is due to a chemico-physical effect: the adsorption to the soil, which holds the contaminant back and does not allow it to progress until the quantity corresponding to the chemical adsorption equilibrium has been adsorbed. This effect is particularly important for less soluble contaminants, which thus can move even hundreds or thousands times slower than water. The effect of this phenomenon is that only more soluble species can cover long distances. The retardation factor depends on the chemical nature of both the contaminant and the aquifer.

GOVERNING EQUATIONS

Darcy's Law

Darcy's law is a Constitutive equation (empirically derived by Henri Darcy, in 1856) that states the amount of ground-water discharging through a given portion of aquifer is proportional to the cross-sectional area of flow, the hydraulic head gradient, and the hydraulic conductivity.

Ground-water Flow Equation

The ground-water flow equation, in its most general form, describes the movement of ground-water in a porous medium (aquifers and aquitards). It is known in mathematics as the diffusion equation, and has many analogs in other fields. Many solutions for ground-water flow problems were borrowed or adapted from existing heat transfer solutions.

It is often derived from a physical basis using Darcy's law and a conservation of mass for a small control volume. The equation is often used to predict flow to wells, which have radial symmetry, so the flow equation is commonly solved in polar or cylindrical coordinates. The Theis equation is one of the most commonly used and fundamental solutions to the ground-water flow equation; it can be used to predict the transient evolution of head due to the effects of pumping one or a number of pumping wells. The Thiem equation is a solution to the steady state ground-water flow equation (Laplace's Equation) for flow to a well. Unless there are large sources of water nearby (a river or lake), true steady-state is rarely achieved in reality.

Both above equations are used in aquifer tests (pump tests). The Hooghoudt equation is a ground-water flow equation applied to subsurface drainage by pipes, tile drains or ditches. An alternative subsurface drainage method is drainage by wells for which ground-water flow equations are also available.

Calculation of Ground-water Flow

To use the ground-water flow equation to estimate the distribution of hydraulic heads, or the direction and rate of ground-water flow, this partial differential equation (PDE) must be solved. The most common means of analytically solving the diffusion equation in the hydrogeology literature are:

- Laplace, Hankel and Fourier transforms (to reduce the number of dimensions of the PDE),
- similarity transform (also called the Boltzmann transform) is commonly how the Theis solution is derived,
- separation of variables, which is more useful for non-Cartesian coordinates, and
- Green's functions, which is another common method for deriving the Theis solution — from the fundamental solution to the diffusion equation in free space.

No matter which method we use to solve the ground-water flow equation, we need both initial conditions (heads at time (t) = 0) and boundary conditions (representing either the physical boundaries of the domain, or an approximation of the domain beyond that point). Often the initial conditions are supplied to a transient simulation, by a corresponding steady-state simulation (where the time derivative in the ground-water flow equation is set equal to 0).

There are two broad categories of how the (PDE) would be solved; either analytical methods, numerical methods, or something possibly in between. Typically, analytic methods solve the ground-water flow equation under a simplified set of conditions *exactly*, while numerical methods solve it under more general conditions to an *approximation*.

Analytic Methods

Analytic methods typically use the structure of mathematics to arrive at a simple, elegant solution, but the required derivation for all but the simplest domain geometries can be quite complex (involving non-standard coordinates, conformal mapping, etc.). Analytic solutions typically are also simply an equation that can give a quick answer based on a few basic parameters. The Theis equation is a very simple (yet still very useful) analytic solution to the ground-water flow equation, typically used to analyze the results of an aquifer test or slug test.

Numerical Methods

The topic of numerical methods is quite large, obviously being of use to most fields of engineering and science in general. Numerical methods have been around much longer than computers have (In the 1920s Richardson developed some of the finite difference schemes still in use today, but they were calculated by hand, using paper and pencil, by human "calculators"), but they have become very important through the availability of fast and cheap personal computers. A quick survey of the main numerical methods used in hydrogeology, and some of the most basic principles is shown below and further discussed in the article "Ground-water model".

There are two broad categories of numerical methods: gridded or discretized methods and non-gridded or mesh-free methods. In the common finite difference method and finite element method (FEM) the domain is completely gridded ("cut" into a grid or mesh of small elements). The analytic element method (AEM) and the boundary integral equation method (BIEM — sometimes also called BEM, or Boundary Element Method) are only discretized at boundaries or along flow elements (line sinks, area sources, etc.), the majority of the domain is mesh-free.

General Properties of Gridded Methods

Gridded Methods like finite difference and finite element methods solve the ground-water flow equation by breaking the problem area (domain) into many small elements (squares, rectangles, triangles, blocks, tetrahedra, etc.) and solving the flow equation for each element (all material properties are assumed constant or possibly linearly variable within an element), then linking together all the elements using conservation of mass across the boundaries between the elements (similar to the divergence theorem). This results in a system which overall approximates the ground-water flow equation, but exactly matches the boundary conditions (the head or flux is specified in the elements which intersect the boundaries).

Finite differences are a way of representing continuous differential operators using discrete intervals (*Äx* and *Ät*), and the finite difference methods are based on these (they are derived from a Taylor series). The forward finite difference approximation is unconditionally stable, but leads to an implicit set of equations (that must be solved using matrix methods, e.g. LU or Cholesky decomposition). The similar backwards difference is only conditionally stable, but it is explicit and can be used to "march" forward in the time direction, solving one grid node at a time (or possibly in parallel, since one node depends only on its immediate neighbours).

Rather than the finite difference method, sometimes the Galerkin FEM approximation is used in space (this is different from the type of FEM often used in structural engineering) with finite differences still used in time.

Application of Finite Difference Models

MODFLOW is a well-known example of a general finite difference ground-water flow model. It is developed by the US Geological Survey as a modular and extensible simulation tool for modeling ground-water flow. It is free software developed, documented and distributed by the USGS. Many commercial products have grown up around it, providing graphical user interfaces to its input file based interface, and typically incorporating pre- and post-processing of user data. Many other models have been developed to work with MODFLOW input and output, making linked models which simulate several hydrologic processes possible (flow and transport models, surface water and ground-water models and chemical reaction models), because of the simple, well documented nature of MODFLOW.

Application of Finite Element Models

Finite Element programs are more flexible in design (triangular elements vs. the block elements most finite difference models use) and there are some programs available (SUTRA, a 2D or 3D density-dependent flow model by the USGS; Hydrus, a commercial unsaturated flow model; FEFLOW, a commercial modelling environment for subsurface flow, solute and heat transport processes; and COMSOL Multiphysics (FEMLAB) a commercial general modelling environment), but unless they are gaining in importance they are still not as popular in with practicing hydrogeologists as MODFLOW is. Finite element models are more popular in university and laboratory environments, where specialized models solve non-standard forms of the flow equation (unsaturated flow, density dependent flow, coupled heat and ground-water flow, etc.)

Application of Finite Volume Models

The finite volume method is a method for representing and evaluating partial differential equations as algebraic equations. Similar to the finite difference method, values are calculated at discrete places on a meshed geometry. "Finite volume" refers to the small volume surrounding each node point on a mesh. In the finite volume method, volume integrals in a partial differential equation that contain a divergence term are converted to surface integrals, using the divergence theorem. These terms are then evaluated as fluxes at the surfaces of each finite volume. Because the flux entering a given volume is identical to that leaving the adjacent volume, these methods are conservative. Another advantage of the finite volume method is that it is easily formulated to allow for unstructured meshes. The method is used in many computational fluid dynamics packages.

The FEHM software package is available free from Los Alamos National Laboratory and can be accessed at the FEHM Website. This versatile porous flow simulator includes capabilities to model multiphase, thermal, stress, and multicomponent reactive chemistry. Current work using this code includes simulation of methane hydrate formation, CO_2 sequestration, oil shale extraction, migration of both nuclear and chemical contaminants, environmental isotope migration in the unsaturated zone, and karst formation.

Other Methods

These include mesh-free methods like the Analytic Element Method (AEM) and the Boundary Element Method (BEM), which are closer to analytic solutions, but they do approximate the ground-water flow equation in some way. The BEM and AEM exactly solve the ground-water flow equation (perfect mass balance), while approximating the boundary conditions. These methods are more exact and can be much more elegant solutions (like analytic methods are), but have not seen as widespread use outside academic and research groups yet.

Causes

The specific contaminants leading to pollution in water include a wide spectrum of chemicals, pathogens, and physical or sensory changes such as elevated temperature and discoloration. While many of the chemicals and substances that are regulated may be naturally occurring (calcium, sodium, iron, manganese, etc.) the concentration is often the key in determining what is a natural component of water, and what is a contaminant. High concentrations of naturally-occurring substances can have negative impacts on aquatic flora and fauna.

Oxygen-depleting substances may be natural materials, such as plant matter (e.g. leaves and grass) as well as man-made chemicals. Other natural and anthropogenic substances may cause turbidity (cloudiness) which blocks light and disrupts plant growth, and clogs the gills of some fish species.

Many of the chemical substances are toxic. Pathogens can produce waterborne diseases in either human or animal hosts. Alteration of water's physical chemistry includes acidity (change in pH), electrical conductivity, temperature, and eutrophication. Eutrophication is an increase in the concentration of chemical nutrients in an ecosystem to an extent that increases in the primary productivity of the ecosystem. Depending on the degree of eutrophication, subsequent negative environmental effects such as anoxia (oxygen depletion) and severe reductions in water quality may occur, affecting fish and other animal populations.

Pathogens

Coliform bacteria are a commonly used bacterial indicator of water pollution, although not an actual cause of disease. Other microorganisms sometimes found in surface waters which have caused human health problems include:

- *Burkholderia pseudomallei*
- *Cryptosporidium parvum*
- *Giardia lamblia*
- *Salmonella*
- *Novovirus* and other viruses
- Parasitic worms (helminths).

High levels of pathogens may result from inadequately treated sewage discharges. This can be caused by a sewage plant designed with less than secondary treatment (more typical in less-developed countries).

In developed countries, older cities with aging infrastructure may have leaky sewage collection systems (pipes, pumps, valves), which can cause sanitary sewer overflows. Some cities also have combined sewers, which may discharge untreated sewage during rain storms.

Pathogen discharges may also be caused by poorly managed livestock operations.

Chemical and other Contaminants

Contaminants may include organic and inorganic substances.

Organic water pollutants include:

- Detergents
- Disinfection by-products found in chemically disinfected drinking water, such as chloroform

- Food processing waste, which can include oxygen-demanding substances, fats and grease
- Insecticides and herbicides, a huge range of organohalides and other chemical compounds
- Petroleum hydrocarbons, including fuels (gasoline, diesel fuel, jet fuels, and fuel oil) and lubricants (motor oil), and fuel combustion byproducts, from stormwater runoff
- Tree and bush debris from logging operations
- Volatile organic compounds (VOCs), such as industrial solvents, from improper storage. Chlorinated solvents, which are dense non-aqueous phase liquids (DNAPLs), may fall to the bottom of reservoirs, since they don't mix well with water and are denser.
- Various chemical compounds found in personal hygiene and cosmetic products.

Inorganic water pollutants include:

- Acidity caused by industrial discharges (especially sulphur dioxide from power plants)
- Ammonia from food processing waste
- Chemical waste as industrial by-products
- Fertilizers containing nutrients—nitrates and phosphates—which are found in stormwater runoff from agriculture, as well as commercial and residential use
- Heavy metals from motor vehicles (via urban stormwater runoff) and acid mine drainage
- Silt (sediment) in runoff from construction sites, logging, slash and burn practices or land clearing sites.

Macroscopic pollution—large visible items polluting the water—may be termed "floatables" in an urban stormwater context, or marine debris when found on the open seas, and can include such items as:

- Trash or garbage (e.g. paper, plastic, or food waste) discarded by people on the ground, along with accidental or intentional dumping of rubbish, that are washed by rainfall into storm drains and eventually discharged into surface waters
- Nurdles, small ubiquitous waterborne plastic pellets
- Shipwrecks, large derelict ships.

Thermal Pollution

Thermal pollution is the rise or fall in the temperature of a natural body of water caused by human influence. Thermal pollution, unlike chemical pollution, results in a change in the physical properties of water. A common cause of thermal pollution is the use of water as a coolant by power plants and industrial manufacturers. Elevated water temperatures decreases oxygen levels (which can kill fish) and affects ecosystem composition, such as invasion by new thermophilic species. Urban runoff may also elevate temperature in surface waters.

Thermal pollution can also be caused by the release of very cold water from the base of reservoirs into warmer rivers.

Transport and Chemical Reactions of Water Pollutants

Most water pollutants are eventually carried by rivers into the oceans. In some areas of the world the influence can be traced hundred miles from the mouth by studies using hydrology transport models. Advanced computer models such as SWMM or the DSSAM Model have been used in many locations worldwide to examine the fate of pollutants in aquatic systems. Indicator filter feeding species such as copepods have also been used to study pollutant fates in the New York Bight, for example.

The highest toxin loads are not directly at the mouth of the Hudson River, but 100 kilometres south, since several days are required for incorporation into planktonic tissue. The Hudson discharge flows south along the coast due to coriolis force. Further south then are areas of oxygen depletion, caused by chemicals using up oxygen and by algae blooms, caused by excess nutrients from algal cell death and decomposition. Fish and shellfish kills have been reported, because toxins climb the food chain after small fish consume copepods, then large fish eat smaller fish, etc. Each successive step up the food chain causes a stepwise concentration of pollutants such as heavy metals (e.g. mercury) and persistent organic pollutants such as DDT. This is known as biomagnification, which is occasionally used interchangeably with bioaccumulation.

Large gyres (vortexes) in the oceans trap floating plastic debris. The North Pacific Gyre for example has collected the so-called "Great Pacific Garbage Patch" that is now estimated at 100 times the size of Texas. Many of these long-lasting pieces wind up in the stomachs of marine birds and animals. This results in obstruction of digestive pathways which leads to reduced appetite or even starvation.

Many chemicals undergo reactive decay or chemically change especially over long periods of time in ground-water reservoirs. A noteworthy class of such chemicals is the chlorinated hydrocarbons such as trichloroethylene (used in industrial metal degreasing and electronics manufacturing) and tetrachloroethylene used in the dry cleaning industry (note latest advances in liquid carbon dioxide in dry cleaning that avoids all use of chemicals). Both of these chemicals, which are carcinogens themselves, undergo partial decomposition reactions, leading to new hazardous chemicals (including dichloroethylene and vinyl chloride).

Ground-water pollution is much more difficult to abate than surface pollution because ground-water can move great distances through unseen aquifers. Non-porous aquifers such as clays partially purify water of bacteria by simple filtration (adsorption and absorption), dilution, and, in some cases, chemical reactions and biological activity: however, in some cases, the pollutants merely transform to soil contaminants. Ground-water that moves through cracks and caverns is not filtered and can be transported as easily as surface water. In fact, this can be aggravated by the human tendency to use natural sinkholes as dumps in areas of Karst topography.

There are a variety of secondary effects stemming not from the original pollutant, but a derivative condition. An example is silt-bearing surface runoff, which can inhibit the penetration of sunlight through the water column, hampering photosynthesis in aquatic plants.

Measurement

Water pollution may be analysed through several broad categories of methods: physical, chemical and biological. Most involve collection of samples, followed by specialized analytical tests. Some methods may be conducted *in situ,* without sampling, such as temperature. Government agencies and research organizations have published standardized, validated analytical test methods to facilitate the comparability of results from disparate testing events.

Sampling

Sampling of water for physical or chemical testing can be done by several methods, depending on the accuracy needed and the characteristics of the contaminant. Many contamination events are sharply restricted in time, most commonly in association with rain events. For this reason "grab" samples are often inadequate for fully quantifying contaminant levels. Scientists gathering this type of data often employ auto-sampler devices that pump increments of water at either time or discharge intervals. Sampling for biological testing involves collection of plants and/or animals from the surface water body. Depending on the type of assessment, the organisms may be identified for biosurveys (population counts) and returned to the water body, or they may be dissected for bioassays to determine toxicity.

Physical Testing

Common physical tests of water include temperature, solids concentration (e.g., total suspended solids (TSS)) and turbidity.

Chemical Testing

Water samples may be examined using the principles of analytical chemistry. Many published test methods are available for both organic and inorganic compounds. Frequently used methods include pH, biochemical oxygen demand (BOD), chemical oxygen demand (COD), nutrients (nitrate and phosphorus compounds), metals (including copper, zinc, cadmium, lead and mercury), oil and grease, total petroleum hydrocarbons (TPH), and pesticides.

Biological Testing

Biological testing involves the use of plant, animal, and/or microbial indicators to monitor the health of an aquatic ecosystem.

Control of Domestic Sewage

Domestic sewage is 99.9 percent pure water, while the other 0.1 percent are pollutants. Although found in low concentrations, these pollutants pose risk on a large scale. In urban areas, domestic sewage is typically treated by centralized sewage treatment plants. In the U.S., most of these plants are operated by local government agencies, frequently referred to as publicly owned treatment works (POTW).

Municipal treatment plants are designed to control conventional pollutants: BOD and suspended solids.

Well-designed and operated systems (i.e., secondary treatment or better) can remove 90 percent or more of these pollutants. Some plants have additional sub-systems to treat nutrients and pathogens. Most municipal plants are not designed to treat toxic pollutants found in industrial wastewater.

Cities with sanitary sewer overflows or combined sewer overflows employ one or more engineering approaches to reduce discharges of untreated sewage, including:

- utilizing a green infrastructure approach to improve stormwater management capacity throughout the system, and reduce the hydraulic overloading of the treatment plant
- repair and replacement of leaking and malfunctioning equipment
- increasing overall hydraulic capacity of the sewage collection system (often a very expensive option).

A household or business not served by a municipal treatment plant may have an individual septic tank, which treats the wastewater on site and discharges into the soil. Alternatively, domestic wastewater may be sent to a nearby privately owned treatment system (e.g. in a rural community).

Industrial Wastewater

Some industrial facilities generate ordinary domestic sewage that can be treated by municipal facilities. Industries that generate wastewater with high concentrations of conventional pollutants (e.g. oil and grease), toxic pollutants (e.g. heavy metals, volatile organic compounds) or other nonconventional pollutants such as ammonia, need specialized treatment systems. Some of these facilities can install a pre-treatment system to remove the toxic components, and then send the partially-treated wastewater to the municipal system. Industries generating large volumes of wastewater typically operate their own complete on-site treatment systems. Some industries have been successful at redesigning their manufacturing processes to reduce or eliminate pollutants, through a process called pollution prevention. Heated water generated by power plants or manufacturing plants may be controlled with:

- cooling ponds, man-made bodies of water designed for cooling by evaporation, convection, and radiation
- cooling towers, which transfer waste heat to the atmosphere through evaporation and/or heat transfer
- cogeneration, a process where waste heat is recycled for domestic and/or industrial heating purposes.

ENVIRONMENTAL WATER POLLUTION

Water pollution is a broad and generic term with a variety of meanings. *It can, in fact, mean almost any type of aquatic contamination between two extremes*:

1. A highly enriched overproductive biotic community, as a lake or river enriched with nitrates and phosphates from domestic sewage;
2. A biotic community with sufficient concentration of toxic substances to eliminate many forms of living organisms or even exclude all forms of life.

Types of water pollution may also be identified by the medium in which they occur (surface water, ground water, soil, etc.), the habitat in which they occur (marine, estuarine, river, etc.), or the source or type of contamination (nutrient, domestic, pesticide, thermal, industrial, etc.). One of the best definitions of pollution is. "Environmental pollution is the unfavorable alteration of our surroundings, wholly or largely as a by-product of man's actions, through direct or indirect effects of changes in energy patterns, radiation levels, chemical and physical constitution and the abundance organisms."

Pollutants of streams, lakes and estuaries come from many sources. Excessive nutrients commonly originate in domestic sewage and run-off from agricultural fertilizer. Certainly the former is the major source of excessive nutrients in most streams and lakes. Toxic chemicals originate in industrial operations, acid waters from mine seepage or surface erosion, and washings of herbicides and insecticides.

Bacterial and Viral Pollution

Water pollution becomes not only an esthetic problem for man, but an endonomic and medical one as well. Bacterial and viral contamination is a threat for the spread of water-borne diseases such as typhoid, shigellosis or bacillary dysentery, amoebic dysentery, cholera and hepatitis. In many watersheds raw sewage is a serious problem. The Hudson River above New York City receives over 200 million gallons per day of raw sewage.

Around many estuarine systems and freshwater lakes and even ocear beaches, public swimming areas have been closed in recent years because of high bacterial counts resulting from domestic pollution. During the summer of 1971, the tourist trade at beach resorts in Belgium, France, Spain and Italy was adversely affected by coastal pollution. Approximately one-half the public bathing beaches of Chesapeake Bay within a twenty mile radius of Baltimore have been closed by public health authorities because of high coliform bacterial counts. These bacteria are taken as indicators of pollution, often, but not exclusively, emanating from domestic sewage.

In Florida, oyster beds in Tampa Bay, Pensacola Bay, Indian River and the St. John's River are closed to harvesting because of domestic pollution. Many parts of San Francisco Bay are closed to water contact sports, as are many beaches of Lake Erie near the larger towns and cities. This does not mean that all these communities live on the brink of typhoid or dysentery epidemics, though the potential does exist in times of flooding, but it does mean that human activities and recreation must be curtailed because of environmental deterioration.

There is still much to be learned about the bacterial communities of polluted aquatic systems. Normally we do not think of serious enteric pathogens being in our water systems in any great abundance. Coliform bacteria, the common indicators of domestic pollution, are not pathogenic in their common forms. The US Public Health Service estimated that a minimum of 40,000 cases of waterborne illness occur every year in the United States. As specific examples, in Riverside and Medera, California, in 1965, waterborne disease caused 20,000 illnesses and several deaths. Approximately 8 million Americans drink water with a bacteriological content exceeding the recommended standards of the USPHS.

Effects of Excessive Nutrients

Even with modern sewage treatment plants, water pollution problems are not entirely avoided. Modern plants remove or inactivate bacteria from the effluent water, but such water is still rich in basic nutrients, such as ammonia, nitrogen, nitrates, nitrites and phosphates. Primary and secondary sewage plants do not remove these sources of pollution. Primary sewage treatment involves screening and sedimentation of solids; secondary treatment involves biological reduction of organic matter; only tertiary treatment removes nutrients. Such nutrients stimulate plant growth, often in the form of phytoplankton or algae.

The enriched waters are thus prone to plankton blooms which may have several undesirable consequences. Some plankton blooms, particularly those of the blue-green algae, produce undesirable odours and tastes in water. Others, such as the dinoflagellate blooms or redtide of the southern coastal regions, produce toxic metabolic products which can result in major fish kills.

Plankton blooms of green algae do not necessarily produce undesirable odours or toxic products, but they can still create problems of oxygen supply in the water. While these blooms exist under abundant sunlight, they contribute oxygen to the water through photosynthesis, but under conditions of continued cloudiness, they consume more oxygen than they produce and lead to oxygen depletion in the waters.

Thus, dissolved oxygen may decline rapidly from favourable levels of 10 to 12 ppm to unfavorable levels of 2 to 3 ppm in which fish experience distress and asphyxiation. Highly enriched streams just below sewage outfalls may show a severe reduction in fish populations, as was documented for the Patuxent River of Maryland. The Potomac River below Washington, DC is highly polluted with domestic sewage and has dissolved oxygen levels often less than 1 ppm. This portion of the river displays annual fish kills every May, when fish reach these oxygen-depleted waters during their spring migrations. Excessive nutrient levels in aquatic systems can cause two other kinds of ecologic consequences. They may lead to extensive growth of aquatic weeds such as Eurasian milfoil, water hyacinth, water chestnut, and many others which have become a worldwide problem. These growths may become so great as to impair fishing, bathing, fish spawning, shellfish production, and even navigation. Hence, excessive plant growths in enriched waters often represent a major economic problem as well as a complete disruption of aquatic ecology.

It has recently been demonstrated that excessive nutrients in water supplies, in the order of 8 or 9 ppm of nitrate nitrogen can cause human disease, as, for example, methemoglobinemia in infants. This is an illness caused by a modified form of normal oxyhemoglobin in the blood, resulting in inadequate oxygen transport by red cells and laboured breathing.

Industrial Pollutants

The effects of industrial wastes on aquatic systems could be the subject of an entire book, and we can do no more here than provide a few examples. In Bellingham Bay and the Straits of Juan de Fuca north of Seattle, Washington, sulfite wastes from pulp mills produce abnormal growth of oyster larvae.

Sometimes the effects of pollution are much more dramatic than the modification of larval growth. In 1959, the US Public Health Service started a national survey of pollution-caused fish kills. In 1961, 45 states reported major fish kills, and in 1962, 38 states reported pollution-caused fish kills totaling 381 separate incident.

Of these 381 fish kills, 43 per cent were attributed to industrial wastes, 13 per cent to agricultural poisons, 8 per cent to domestic sewage, 5 per cent to mining operations and 31 per cent were of unknown origin.

Some of the various conditions or accidents which caused these kills were as follows:

- Chemicals were used to flush the lines of a power plant (sodium nitrite, hydrazine, ammonium bifluoride, and soda ash);
- Bags of endrin used for insect control were accidentally dumped into a stream;
- The unauthorized application of copper sulfate was used to control a plankton bloom;
- Outfall from a paper mill contained lignite and sulfite waste liquor;
- Acid mine drainage discharged into a stream;
- Concentrated sulfuric acid released from a cotton seed delinting plant;
- Seepage of the pesticides chlordane and heptachlor from termite treatment project;
- Chromium solution from a plating tank leaked into a sewer system;
- Cyanide released from drains of blast furnaces at a steel company;
- Hot water from a steam generating plant released into a stream.

These are merely representative examples of the types of toxic effects which industrial and agricultural operations may have unless rigorous control measures are maintained. Increasing use of powerful chemicals in industrial and agricultural operations increases the risk of environmental damage through accidents. Occasionally fish kills involve massive populations of aquatic life. In 1962 in San Diego harbour, an estimated

37,800,000 fish were killed by pollution, producing one raft of dead fish 1000 feet long, 10 feet wide and 3 feet deep. In 1963, an extensive fish kill in Chesapeake Bay killed millions of fish from Baltimore to Norfolk, and in 1967 many acres of dead alewife fish washed up along the Lake Michigan shores of the Chicago waterfront. The Chesapeake Bay and Lake Michigan fish kills may have been the result of epizootics of infectious disease, however, and were not necessarily the direct result of pollution. Certainly pollution played a role in these outbreaks, however, and may have been a triggering factor.

Another major problem in water pollution is the addition of various ions and chemicals in the water which have toxic effects on plant, animal and human life. Chlorine added to water to control bacteria and algae may persist in streams to cause mortality of plankton and fish. This occurred in Maryland in chlorine-treated waters used to cool an electric power plant on the Patuxent River. Mercury, as a by-product of industrial operations involved in the production of vinyl chloride, has cropped up as a toxic agent of serious proportions.

Mercury is used in many chemical industries, and it also emerges as a by-product of some incinerators, power plants, laboratories, and even hospitals, to the extent of 23 million pounds per year throughout the world. In Japan, human illness and death occurred in the 1950s among fishermen who ingested fish, crabs and shellfish contaminated with a simple alkyl mercury compound from Japanese coastal industries. This mercury poisoning produced a crippling and often fatal disease known as Minamata disease.

The name was derived from Minamata Bay on the southwest coast of Kyushu. Minamata disease was characterized by substantial pathology of the central nervous system and voluntary musculature. Initial symptoms included numbness of the limbs; lips and tongue, impairment of motor control, deafness, and blurring of vision. Cellular degeneration occurred in the cerebellum, midbrain, and cerebral cortex, and this led to spasticity, rigidity, stupor, and coma.

The first cases of Minamata disease appeared in Japan in 1953, and of 52 original patients in the villages around Minamata Bay, 17 died, 23 were permanently disabled, and only 3 were able to return to work within 6 months. Minamata disease has not occurred in the United States as a result of eating coastal fish, but in 1969 it was discovered that fish in Lake St. Clair and Lake Erie contained mercury above permissible levels. The Canadian government immediately banned the commercial sale of fish from these waters until the industrial sources of mercury were removed or corrected, and the mercury levels of fish fell below critical limits. In 1971, the US Food and Drug Administration banned the consumption of tinned swordfish because 80 per cent of the samples tested had excessive levels of mercury. Some samples of tuna fish have also shown high levels of mercury, and in 1971 a New York woman who had been dieting by eating tuna every day developed mercury poisoning.

Mercury is just one of many toxic chemicals which may occur in water polluted by industrial or agricultural effluents, Lead, cadmium, and nickel carbonyl are other pollutants which are toxic or pethogenic for man and animals. In fact, more than 12,000 toxic industrial chemicals are in use today. With over one-half of our total available fresh water row being used, one can glimpse the immense ecologic problem confronting modern society.

A basic concern in the overall picture of water pollution is the finite nature of our present surface and ground water supplies. Water use for domestic and industrial purposes is increasing so dramatically, that the time when all available fresh water will be used is now in sight for many communities. Hydrologists estimate the total fresh water supplies of the United States at 515 billion gallons per day, of which we are already using 360 billion gallons per day.

The metropolitan areas of Philadelphia and Wilmington are now using nearly 80 per cent of their total available water. This shall reach 100 per cent within 10 years at present consumption trends. The great hope, of course, lies in finding new sources and in desalination, but these solutions contain economic and technological problems that remain to be solved. Water pollution and water supply are intimately related because of the necessity for and economics of reuse.

EUTROPHICATION

Eutrophication is caused by the decrease of an ecosystem with chemical nutrients, typically compounds containing nitrogen or phosphorus. It may occur on land or in the water. Eutrophication is frequently a result of nutrient pollution such as the release of sewage effluent into natural waters (rivers or coasts) although it may occur naturally in situations where nutrients accumulate (e.g. depositional environments) or where they flow into systems on an ephemeral basis (e.g. intermittent upwelling in coastal systems). Eutrophication generally promotes excessive plant growth and decay, favours certain weedy species over others, and is likely to cause severe reductions in water quality. In aquatic environments, enhanced growth of choking aquatic vegetation or phytoplankton (that is, an algal bloom) disrupts normal functioning of the ecosystem, causing a variety of problems. Human society is impacted as well: eutrophication decreases the resource value of rivers, lakes, and estuaries such that recreation, fishing, hunting, and aesthetic enjoyment are hindered. Health-related problems can occur where eutrophic conditions interfere with drinking water treatment. Although traditionally thought of as enrichment of aquatic systems by addition of fertilizers into lakes, bays, or other semi-enclosed waters (even slow-moving rivers), terrestrial ecosystems are subject to similarly adverse impacts. Increased content of nitrates in soil frequently leads to undesirable changes in vegetation composition and many plant species are endangered as a result of eutrophication in terrestric ecosystems, e.g. majority of orchid species in Europe. Ecosystems (like some meadows, forests and bogs that are characterized by low nutrient content and species-rich, slowly growing vegetation adapted to lower nutrient levels) are overgrown by faster growing and more competitive species-poor vegetation, like tall grasses, that can take advantage of unnaturally elevated nitrogen level and the area may be changed beyond recognition and vulnerable species may be lost. Eg. species-rich fens are overtaken by reed or reedgrass species, spectacular forest undergrowth affected by run-off from nearby fertilized field is turned into a thick nettle and bramble shrub.

Eutrophication was recognized as a pollution problem in European and North American lakes and reservoirs in the mid-20th century. Since then, it has become more widespread. Surveys showed that 54% of lakes in Asia are eutrophic; in Europe, 53%; in North America, 48%; in South America, 41%; and in Africa, 28%.

Concept of Eutrophication

Eutrophication can be a natural process in lakes, as they fill in through geological time, though other lakes are known to demonstrate the reverse process, becoming less nutrient rich with time. Estuaries also tend to be naturally eutrophic because land-derived nutrients are concentrated where run-off enters the marine environment in a confined channel and mixing of relatively high nutrient fresh water with low nutrient marine water occurs.

Phosphorus is often regarded as the main culprit in cases of eutrophication in lakes subjected to point source pollution from sewage. The concentration of algae and the tropic state of lakes correspond well to phosphorus levels in water. Studies conducted in the Experimental Lakes Area in Ontario have shown a relationship between the addition of phosphorus and the rate of eutrophication. Humankind has increased the rate of phosphorus cycling on Earth by four times, mainly due to agricultural fertilizer production and application. Between 1950 and 1995, 600,000,000 tonnes of phosphorus were applied to Earth's surface, primarily on croplands. Control of point sources of phosphorus have resulted in rapid control of eutrophication, mainly due to policy changes. Human activities can accelerate the rate at which nutrients enter ecosystems. Runoff from agriculture and development, pollution from septic systems and sewers, and other human-related activities increase the flux of both inorganic nutrients and organic substances into terrestrial, aquatic, and coastal marine ecosystems (including coral reefs). Elevated atmospheric compounds of nitrogen can increase soil nitrogen availability.

Chemical forms of nitrogen are most often of concern with regard to eutrophication because plants have high nitrogen requirements so that additions of nitrogen compounds stimulate plant growth (primary production). Nitrogen is not readily available in soil because N_2, a gaseous form of nitrogen, is very stable and unavailable directly to higher plants. Terrestrial ecosystems rely on microbial nitrogen fixation to convert N_2 into other physical forms (such as nitrates). However, there is a limit to how much nitrogen can be utilized. Ecosystems receiving more nitrogen than the plants require are called nitrogen-saturated. Saturated terrestrial ecosystems contribute both inorganic and organic nitrogen to freshwater, coastal, and marine eutrophication, where nitrogen is also typically a limiting nutrient. However, in marine environments, phosphorus may be limiting because it is leached from the soil at a much slower rate than nitrogen, which are highly insoluble.

Ecological Effects

Adverse effects of eutrophication on lakes, reservoirs, rivers and coastal marine waters. Many ecological effects can arise from stimulating primary production, but there are three particularly troubling ecological impacts: decreased biodiversity, changes in species composition and dominance, and toxicity effects.

Decreased Biodiversity

When an ecosystem experiences an increase in nutrients, primary producers reap the benefits first. In aquatic ecosystems, species such as algae experience a population increase (called an algal bloom). Algal blooms limit the sunlight available to bottom-dwelling organisms and cause wide swings in the amount of dissolved oxygen in the water.

Oxygen is required by all respiring plants and animals and it is replenished in daylight by photosynthesizing plants and algae. Under eutrophic conditions, dissolved oxygen greatly increases during the day, but is greatly reduced after dark by the respiring algae and by microorganisms that feed on the increasing mass of dead algae. When dissolved oxygen levels decline to hypoxic levels, fish and other marine animals suffocate. As a result, creatures such as fish, shrimp, and especially immobile bottom dwellers die off. In extreme cases, anaerobic conditions ensue, promoting growth of bacteria such as Clostridium botulinum that produces toxins deadly to birds and mammals. Zones where this occurs are known as dead zones.

New Species Invasion

Eutrophication may cause competitive release by making abundant a normally limiting nutrient. This process causes shifts in the species composition of ecosystems. For instance, an increase in nitrogen might allow new, competitive species to invade and outcompete original inhabitant species. This has been shown to occur in New England salt marshes.

Toxicity

Some algal blooms, otherwise called "nuisance algae" or "harmful algal blooms," are toxic to plants and animals. Toxic compounds they produce can make their way up the food chain, resulting in animal mortality. Freshwater algal blooms can pose a threat to livestock. When the algae die or are eaten, neuro-and hepatotoxins are released which can kill animals and may pose a threat to humans. An example of algal toxins working their way into humans is the case of shellfish poisoning. Biotoxins created during algal blooms are taken up by shellfish (mussels, oysters), leading to these human foods acquiring the toxicity and poisoning humans. Examples include paralytic, neurotoxic, and diarrhoetic shellfish poisoning. Other marine animals can be vectors for such toxins, as in the case of ciguatera, where it is typically a predator fish that accumulates the toxin and then poisons humans. Nitrogen can also cause toxic effects directly. When this nutrient is leached into groundwater, drinking water can be affected because concentrations of nitrogen are not filtered out. Nitrate (NO_3) has been shown to be toxic to human babies. This is because bacteria can live in their digestive tract that convert nitrate to nitrite (NO_2). Nitrite reacts with hemoglobin to form methemoglobin, a form that does not carry oxygen. The baby essentially suffocates as its body receives insufficient oxygen.

Sources of High Nutrient Runoff

Characteristics of point and nonpoint sources of chemical inputs. In order to gauge how to best prevent eutrophication from occurring, specific sources that contribute to nutrient loading must be identified. There are two common sources of nutrients and organic matter: point and nonpoint sources.

Point Sources

Point sources are directly attributable to one influence. In point sources the nutrient waste travels directly from source to water. For example, factories that have waste discharge pipes directly leading into a water body would be classified as a point source. Point sources are relatively easy to regulate.

Nonpoint Sources

Nonpoint source pollution (also known as 'diffuse' or 'runoff' pollution) is that which comes from ill-defined and diffuse sources. Nonpoint sources are difficult to regulate and usually vary spatially and temporally (with season, precipitation, and other irregular events). It has been shown that nitrogen transport is correlated with various indices of human activity in watersheds, including the amount of development. Agriculture and development are activities that contribute most to nutrient loading. There are three reasons that nonpoint sources are especially troublesome:

Soil Retention

Nutrients from human activities tend to accumulate in soils and remain there for years. It has been shown that the amount of phosphorus lost to surface waters increases linearly with the amount of phosphorus in the soil. Thus much of the nutrient loading in soil eventually makes its way to water. Nitrogen, similarly, has a turnover time of decades or more.

Runoff to Surface Water and Leaching to Groundwater

Nutrients from human activities tend to travel from land to either surface or ground water. Nitrogen in particular is removed through storm drains, sewage pipes, and other forms of surface runoff. Nutrient losses in runoff and leachate are often associated with agriculture. Modern agriculture often involves the application of nutrients onto fields in order to maximise production. However, farmers frequently apply more nutrients than are taken up by crops or pastures. Regulations aimed at minimising nutrient exports from agriculture are typically far less stringent than those placed on sewage treatment plants and other point source polluters.

Atmospheric Deposition

Nitrogen is released into the air because of ammonia volatilization and nitrous oxide production. The combustion of fossil fuels is a large human-initiated contributor to atmospheric nitrogen pollution. Atmospheric deposition (e.g., in the form of acid rain) can also effect nutrient concentration in water, especially in highly industrialized regions.

Other Causes

Any factor that causes increased nutrient concentrations can potentially lead to eutrophication. In modeling eutrophication, the rate of water renewal plays a critical role; stagnant water is allowed to collect more nutrients than bodies with replenished water supplies. It has also been shown that the drying of wetlands causes an increase in nutrient concentration and subsequent eutrophication booms.

Prevention and Reversal

Eutrophication poses a problem not only to ecosystems, but to humans as well. Reducing eutrophication should be a key concern when considering future policy, and a sustainable solution for everyone, including farmers and ranchers, seems feasible. While eutrophication does pose problems, humans should be aware that natural runoff (which causes algal blooms in the wild) is common in ecosystems and should thus not reverse nutrient concentrations beyond normal levels.

Effectiveness

Cleanup measures have been mostly, but not completely, successful. Finnish phosphorus removal measures started in the mid-1970s and have targeted rivers and lakes polluted by industrial and municipal discharges. These efforts have had a 90% removal efficiency. Still, some targeted point sources did not show a decrease in runoff despite reduction efforts.

Minimizing Nonpoint Pollution: Future Work

Nonpoint pollution is the most difficult source of nutrients to manage. The literature suggests, though, that when these sources are controlled, eutrophication decreases. The following steps are recommended to minimize the amount of pollution that can enter aquatic ecosystems from ambiguous sources.

Riparian Buffer Zones

Studies show that intercepting non-point pollution between the source and the water is a successful mean of prevention. Riparian buffer zones, an interface between a flowing body of water and land, have been created near waterways in an attempt to filter pollutants; sediment and nutrients are deposited here instead of in water. Creating buffer zones near farms and roads is another possible way to prevent nutrients from traveling too far. Still, studies have shown that the effects of atmospheric nitrogen pollution can reach far past the buffer zone. This suggests that the most effective means of prevention is from the primary source.

Prevention Policy

Laws regulating the discharge and treatment of sewage have led to dramatic nutrient reductions to surrounding ecosystems, but it is generally agreed that a policy regulating agricultural use of fertilizer and animal waste must be imposed. In Japan the amount of nitrogen produced by livestock is adequate to serve the fertilizer needs for the agriculture industry. Thus, it is not unreasonable to command livestock owners to clean up animal waste-which when left stagnant will leach into ground water.

Nitrogen Testing and Modeling

Soil Nitrogen Testing (N-Testing) is a technique that helps farmers optimize the amount of fertilizer applied to crops. By testing fields with this method, farmers saw a decrease in fertilizer application costs, a decrease in nitrogen lost to surrounding sources, or both. By testing the soil and modeling the bare minimum amount of fertilizer needed, farmers reap economic benefits while the environment remains clean.

Natural State of Algal Blooms

Although the intensity, frequency and extent of algal blooms has tended to increase in response to human activity and human-induced eutrophication, algal blooms are a naturally-occurring phenomenon. The rise and fall of algae populations, as with the population of other living things, is a feature of a healthy ecosystem. Rectification actions aimed at abating eutrophication and algal blooms are usually desirable, but the focus of intervention should not necessarily be aimed at eliminating blooms, but towards creating a sustainable balance that maintains or improves ecosystem health.

6

Irrigation

Irrigation may be defined as the science of artificial application of water to the land or soil. It is used to assist in the growing of agricultural crops, maintenance of landscapes, and revegetation of disturbed soils in dry areas and during periods of inadequate rainfall. Additionally, irrigation also has a few other uses in crop production, which include protecting plants against frost, suppressing weed growing in grain fields and helping in preventing soil consolidation. In contrast, agriculture that relies only on direct rainfall is referred to as rain-fed or dryland farming. Irrigation systems are also used for dust suppression, disposal of sewage, and in mining. Irrigation is often studied together with drainage, which is the natural or artificial removal of surface and sub-surface water from a given area.

Irrigation is also a term used in medical/dental fields to refer to flushing and washing out anything with water or another liquid.

TYPES OF IRRIGATION

Various types of irrigation techniques differ in how the water obtained from the source is distributed within the field. In general, the goal is to supply the entire field uniformly with water, so that each plant has the amount of water it needs, neither too much nor too little.The modern methods are efficient enough to achieve this goal.

Surface

In surface irrigation systems, water moves over and across the land by simple gravity flow in order to wet it and to infiltrate into the soil. Surface irrigation can be subdivided into furrow, *borderstrip or basin irrigation*. It is often called flood irrigation when the irrigation results in flooding or near flooding of the cultivated land. Historically, this has been the most common method of irrigating agricultural land. Where water levels from the irrigation source permit, the levels are controlled by dikes, usually plugged by soil.

This is often seen in terraced rice fields (rice paddies), where the method is used to flood or control the level of water in each distinct field. In some cases, the water is pumped, or lifted by human or animal power to the level of the land.

Localized

Localized irrigation is a system where water is distributed under low pressure through a piped network, in a pre-determined pattern, and applied as a small discharge to each plant or adjacent to it. Drip irrigation, spray or micro-sprinkler irrigation and bubbler irrigation belong to this category of irrigation methods.

Drip

Drip irrigation, also known as trickle irrigation, functions as its name suggests.In this system water falls drop by drop just at the position of roots.

Water is delivered at or near the root zone of plants, drop by drop. This method can be the most water-efficient method of irrigation, if managed properly, since evaporation and run-off are minimized.

In modern agriculture, drip irrigation is often combined with plastic mulch, further reducing evaporation, and is also the means of delivery of fertilizer. The process is known as *fertigation*.

Deep percolation, where water moves below the root zone, can occur if a drip system is operated for too long or if the delivery rate is too high. Drip irrigation methods range from very high-tech and computerized to low-tech and labour-intensive.

Lower water pressures are usually needed than for most other types of systems, with the exception of low energy center pivot systems and surface irrigation systems, and the system can be designed for uniformity throughout a field or for precise water delivery to individual plants in a landscape containing a mix of plant species.

Although it is difficult to regulate pressure on steep slopes, pressure compensating emitters are available, so the field does not have to be level. High-tech solutions involve precisely calibrated emitters located along lines of tubing that extend from a computerized set of valves.

Sprinkler

In sprinkler or overhead irrigation, water is piped to one or more central locations within the field and distributed by overhead high-pressure sprinklers or guns. A system utilizing sprinklers, sprays, or guns mounted overhead on permanently installed risers is often referred to as a *solid-set* irrigation system.

Higher pressure sprinklers that rotate are called *rotors* and are driven by a ball drive, gear drive, or impact mechanism. Rotors can be designed to rotate in a full or partial circle. Guns are similar to rotors, except that they generally operate at very high pressures of 40 to 130 lbf/in^2 (275 to 900 kPa) and flows of 50 to 1200 US gal/min (3 to 76 L/s), usually with nozzle diameters in the range of 0.5 to 1.9 inches (10 to 50 mm). Guns are used not only for irrigation, but also for industrial applications such as dust suppression and logging.

Sprinklers can also be mounted on moving platforms connected to the water source by a hose. Automatically moving wheeled systems known as *traveling sprinklers* may irrigate areas such as small farms, sports fields, parks, pastures, and cemeteries unattended. Most of these utilize a length of polyethylene tubing wound on a steel drum. As the tubing is wound on the drum powered by the irrigation water or a small gas engine, the sprinkler is pulled across the field. When the sprinkler arrives back at the reel the system shuts off. This type of system is known to most people as a "waterreel" traveling irrigation sprinkler and they are used extensively for dust suppression, irrigation, and land application of waste water. Other travellers use a flat rubber hose that is dragged along behind while the sprinkler platform is pulled by a cable. These cable-type travellers are definitely old technology and their use is limited in today's modern irrigation projects.

Center Pivot

Center pivot irrigation is a form of sprinkler irrigation consisting of several segments of pipe (usually galvanized steel or aluminum) joined together and supported by trusses, mounted on wheeled towers with sprinklers positioned along its length. The system moves in a circular pattern and is fed with water from the pivot point at the center of the arc. These systems are found and used in all parts of the world and allow irrigation of all types of terrain. Newer systems have drop sprinkler heads as shown in the image that follows. Most center pivot systems now have drops hanging from a u-shaped pipe attached at the top of the pipe with sprinkler heads that are positioned a few feet (at most) above the crop, thus limiting evaporative losses. Drops can also be used with drag hoses or bubblers that deposit the water directly on the ground between crops. Crops are often planted in a circle to conform to the center pivot. This type of system is known as LEPA (Low Energy Precision Application). Originally, most center pivots were water powered. These were replaced by hydraulic systems (*T-L Irrigation*) and electric motor driven systems (Reinke, Valley, Zimmatic). Many modern pivots feature GPS devices.

Lateral Move (Side Roll, Wheel Line)

A series of pipes, each with a wheel of about 1.5 m diameter permanently affixed to its midpoint and sprinklers along its length, are coupled together at one edge of a field. Water is supplied at one end using a large hose. After sufficient water has been applied, the hose is removed and the remaining assembly rotated either by hand or with a purpose-built mechanism, so that the sprinklers move 10 m across the field. The hose is reconnected.

The process is repeated until the opposite edge of the field is reached. This system is less expensive to install than a center pivot, but much more labour intensive to operate, and it is limited in the amount of water it can carry. Most systems utilize 4 or 5-inch (130 mm) diameter aluminum pipe. One feature of a lateral move system is that it consists of sections that can be easily disconnected. They are most often used for small or oddly shaped fields, such as those found in hilly or mountainous regions, or in regions where labour is inexpensive.

Subirrigation also sometimes called *seepage irrigation* has been used for many years in field crops in areas with high water tables. It is a method of artificially raising the water table to allow the soil to be moistened from below the plants' root zone. Often those systems are located on permanent grasslands in lowlands or river valleys and combined with drainage infrastructure. A system of pumping stations, canals, weirs and gates allows it to increase or decrease the water level in a network of ditches and thereby control the water table.

Sub-irrigation is also used in commercial greenhouse production, usually for potted plants. Water is delivered from below, absorbed upwards, and the excess collected for recycling.

Typically, a solution of water and nutrients floods a container or flows through a trough for a short period of time, 10–20 minutes, and is then pumped back into a holding tank for reuse. Sub-irrigation in greenhouses requires fairly sophisticated, expensive equipment and management. Advantages are water and nutrient conservation, and labour-saving through lowered system maintenance and automation. It is similar in principle and action to subsurface drip irrigation.

Manual using Buckets or Watering Cans

These systems have low requirements for infrastructure and technical equipment but need high labour inputs. Irrigation using watering cans is to be found for example in peri-urban agriculture around large cities in some African countries.

Automatic, Non-electric using Buckets and Ropes

Besides the common manual watering by bucket, an automated, natural version of this also exist. Using plain polyester ropes combined with a prepared ground mixture can be used to water plants from a vessel filled with water.

The ground mixture would need to be made depending on the plant itself, yet would mostly consist of black potting soil, vermiculite and perlite. This system would (with certain crops) allow to save expenses as it does not consume any electricity and only little water (unlike sprinklers, water timers,...). However, it may only be used with certain crops (probably mostly larger crops that do not need a humid environment; perhaps *e.g.* paprikas).

Using Water Condensed from Humid Air

In countries where at night, humid air sweeps the countryside, water can be obtained from the humid air by condensation onto cold surfaces. This is for example practiced in the vineyards at Lanzarote using stones to condense water or with various fog collectors based on canvas or foil sheets.

IMPORTANCE OF IRRIGATION SYSTEM

John Skutsch and Darren Evans have made a valuable contribution to the debate on irrigation maintenance and this section draws heavily on their work. They suggest that widespread evidence points to a considerable shortfall between recommended maintenance expenditure for public-sector irrigation schemes and the amounts actually spent. In a variety of projects researched during the 1990s in India, Indonesia, Pakistan and Sri Lanka, the ratio per hectare of actual spend to recommended spend was only 24-47 per cent.

In the first years of a new scheme, maintenance neglect may have no dramatic effect. Irrigation staff can freeride on the freeboard produced by the vertical distance by which the channel flow's usual height exceeds the required operating level. Cropping systems, too, may show some resilience. But sooner or later maintenance underfunding begins to bite. Schemes in arid and semi-arid areas, where crops are entirely dependent on irrigation under managed rotations, are more sensitive to neglect than projects in the humid tropics that have paddy rice under continuous irrigation supplementary to monsoon rains.

The biggest losses come through crop output decline as the volume of irrigation water delivered eventually begins to decline. This may show itself as a fall in tonnage efficiency or in a diminution of the irrigated area. Moreover, as Carruthers and Morrison indicate, farmers shift to lower value crops so as to be able to reduce risk by limiting the use of inputs. Impeded drainage can lead to waterlogging and salinity and to an increase in the incidence of water-related diseases associated with blocked channels and stagnant water. Skutsch and Evans, commenting on the serious social and financial implications for farmers, write:

Those in the more favoured parts of the system may continue to receive an adequate water supply, while those at the tail-end can face ruin. Poor maintenance thus directly aggravates existing inequities within the farming community. It initiates a vicious circle of decline: reduced water supply; lost output; farmers' anger, despair and reduced investment; reduced water fee collections; vandalism and conflict. Smallholder farmers operate on a narrow margin between relative success and failure. Land is often mortgaged against the following harvest to pay for the cost of agricultural inputs. A single disastrous cropping season can mean the loss of a farmer's land and enforced migration to the cities to seek work.

This process brings with it the premature obsolescence of irrigation projects. The search for new capital funding begins in order to renovate the assets far earlier than would have been necessary under a satisfactory maintenance regime. A cycle of build-neglect-rebuild is in place. Partly as a result of this, most development bank finance for irrigation is now for the rehabilitation of existing schemes. Skutsch estimates that about two-thirds of recent international lending has been for systems that have suffered early technical failure. Already in 1995 Jones's overview of World Bank lending for irrigation projects could conclude: 'O&M problems can be seen in the Bank's financing of so many rehabilitation projects. Almost all of them, when scrutinized, turn out to be deferred maintenance projects.

Why, then, is maintenance neglected? As so often is the case in the social sciences, the question is simple but the answer is fiendishly complex. In part, farmers themselves are responsible. As is commonly the case with public-sector irrigation schemes, they see maintenance as the proper task of government.

If farmers are required to allocate greater resources to maintenance, either in terms of increased payments or of increased responsibility...they must see a clear incentive in better cash returns, reduced costs or improved control over water. Particularly on rice-growing schemes in monsoon climates, farmers may see little urgency in paying today to avoid increasing problems tomorrow, particularly if the government has traditionally stepped in to rehabilitate defective systems. In part, government and irrigation department staff are responsible for the neglect of maintenance.

Too often, national, regional and local élites regard the public sector as a milch cow whose teats are there to be squeezed for easy profits. No simpler way exists of doing this than by pushing superfluous staff into administrative posts as a favour to family, professional or political allies. As a consequence the 'establishment' of the department becomes excessively large, draining resources from OM&M. Moreover, operations activities in any case enjoy precedence over maintenance.

Technical and engineering staff, trained in design and construction, are likely to be unmotivated by the bread-and-butter tasks of OM&M, which offer poor rewards and prospects. For senior staff, administrators and engineers, new projects are likely to be the most highly sought after given their prestige and the rent-skimming opportunities they offer. A rule of thumb of some international consultants in parts of the developing world is that capital rather than current spending is favoured by country élites in government because 40 per cent of capital costs go in pay-offs.

At a level more mundane but of real importance, irrigation departments commonly remit the charges they collect from farmers to the finance ministry. That ministry may give other sectors more support than agriculture when it provides revenue financing. As the sum allocated to OM&M is decoupled from the fees collected from farmers, irrigation agencies have no incentive to maximize their revenue collection. There is a parallel here with the breaking of the line in prudential responsibility, itself a 'decoupling' of loan provider and loan user. In part, the development banks themselves are responsible for maintenance neglect. If they lend money on schemes that 15 years later are choked with weeds, silted up, waterlogged, saline, exhibiting high distribution losses and with defective equipment, then their original cost-benefit analyses must have been *seriously* erroneous. The McNamara effect and a weak record in post-project evaluation are partially to blame here. The development banks need not fear that their loans will never be repaid, because the dollar flow of principal and interest comes from national government treasuries, not from the deficient project.

Several authors have proposed policy changes aimed at ending the buildneglect-rebuild syndrome. Such proposals include the following:

- An information system should be put in place that is capable of defining in detail the volume of water used.
- A dependable delivery system should be achieved.
- An agency willing and able to collect fees should be built, with a transparent management system.
- Farmers should be required to contribute an agreed number of maintenance days each year.
- Fees paid by farmers should increase but only after
  - The managing agency details all costs, levels of service and benefits,

  - Users are surveyed on their willingness and ability to pay,
  - The charging mechanism is agreed,
  - Users participate in setting the levels of service
  - Fees are linked to the level of service.
- Recognized sanctions for non-payment should be enforced.
- Feeding back the charges paid by farmers directly to OM&M.
- Maintenance should be carried out by private or autonomous self-accounting agencies.
- Engineers and planners should design irrigation projects for low maintenance, less costly management and greater effective lives.
- Improvement and modernization projects would be considered only for those schemes where a regular review of their individual maintenance standard shows them to be satisfactory or better.
- Lending agencies should provide financial support for a transitional period following construction.
- Tradeable abstraction rights should be introduced.
- Irrigation management should be transferred to farmers themselves.

Managing Common Property Resources

The above list of thirteen proposals for ending the build-neglect-rebuild cycle ends with irrigation management transfer from the public sector to farmers themselves. In the 1990s of such a possibility was able to draw upon and contribute to, a vigorous literature exploring the subject of common property resource management. The published papers and books demonstrate a welcome (if rare) willingness of economists and sociologists to work together in common cause. Below, synthesis of the relevant literature, particularly as it applies to the management of irrigation systems.

In every society common property resources exist, each of which is the common location of work by human agents or actors-the two terms are deployed here interchangeably. Agents may be individual persons, families or institutions. Such common property resources may be constituted by specific natural environments, such as a commons used for grazing livestock, or by some complex resource integrating features of the natural environment with the means of production created by human society, such as an irrigation scheme. The 'common property' characteristic of the resource is that a number of separate actors enjoy rights of access and use of the resource-rights that are recognized in law or are customary practice.

In the day-to-day work located on the common resource, each agent may be motivated only by the *private* interest of that person, family or institution. This is of special importance in three ways. *First,* each actor may seek to appropriate as much of the resource for itself, without regard for other agents' interests. *Second,* each actor may choose to maintain the productivity of the resource only to the degree that the private costs borne by the actor in such efforts are outweighed by the private advantages, without regard for other agents' interests. *Third,* insofar as the use of the common resource creates negative side-effects, the actor may seek to reduce these side-effects only to the extent that their cost outweighs their production advantage, without regard for other agents' interests. So, for any single actor, private interest may prevail over the common interest in respect of resource appropriation, resource maintenance and resource degradation.

Production of outputs from a common resource-the milk yield of ruminants or the tonnage of grain-may grow vigorously over time. Work motivated by the private interests of the agent can stimulate long hours in the field, a sharp tactical appreciation of gains and losses from adaptive action and an eager search for innovative practices. In respect of the productivity of the common resource, these are powerful advantages.

However, with the passage of time the common property resource itself and the patterns of productive activity on it may show signs of enfeeblement and impending long-term collapse. Now, the disadvantages imposed by each single agent's actions on other persons, families and institutions have become disproportionately large. The struggle by actors over the appropriation of the resource can weaken friendship, create mistrust and result in disabling and destructive legal and physical conflicts as well as the exhaustion of the resource itself. Maintenance and protection of the common resource may decline below what is required for its long-term sustainability. Each agent's indifference to the negative effects of its actions on others may bring with it both bitter social disputes as well as the poisoning of the common resource.

Such a trajectory may lead to demands for a reconstitution of the social relationships between actors in the management of the common property resource. Where government cannot or will not lead this collective activity, a non-governmental local entity for collective action (LECA) may be created. The core objectives of the LECA are likely to be fourfold. *First,* the shared use of the resource should be recognized by the participant agents as being equitable. *Second,* the maintenance of the resource should be adequate to ensure its long-term viability. *Third,* the production of negative externalities should be sufficiently well regulated that their collective cost to the new institution's actors is acceptable and does not threaten systemic sustainability. *Fourth,* the transaction costs of meeting the first three objectives are acceptable to actors in the light of the benefits they bring.

There is a problem. A successful LECA brings all-round advantages to its actor-members. But every agent knows that if *he and he alone* continues to pursue his private interest, he reaps the benefits of unrestrained action as well as the benefits deriving from the collective agreement. This is known as the freerider problem. As freerider numbers grow, the advantages of collective action diminish and the LECA collapses.

So each LECA needs to engage in forms of moral persuasion in order that its actors honour the collective agreement made. It will also monitor members' activities to ensure that they do not breach the rules and that, when this occurs, sanctions are exercised against such infractions. The collective also faces costs of bargaining, contract formulation and information search in pursuit of a collective economic strategy. Together, all these costs of motivation, control and co-ordination are termed transaction costs. Even in a study as exhaustive as Anna Blomqvist's in South India it was never possible to assign these costs a monetary value. The greater the LECA's legitimacy, *i.e.* the greater the internalization of its values and ideas within and outside the institution, the less costly is monitoring and enforcement. Blomqvist concluded from her experience that LECAs are most likely to succeed when the members are culturally homogeneous, small in number, highly dependent on the common resource and benefit equally from collective action.

Abstraction charges

Rivers, lakes and aquifers are common property resources, except where access to them is privatized. This is why common property resource analysis, including the appraisal of transaction costs, is of such great relevance to a full understanding of the management of water resources. In particular, rivers, lakes and aquifers are resources from which water is abstracted for use in agriculture, industry and households. They are also resources used as sinks for the disposal of waste water and drainage water.

Abstraction usually takes place where the abstracting actor enjoys formal legal rights to do so, or where abstraction is an accepted customary practice. In such cases the abstractor may be required to pay a charge either to a public agency or to a private owner of rights in water. Abstraction rights may also be traded. One objective of these two sections is to provide a general economic analysis of the subject capable of international application to specific case studies. The second objective is to recommend the most appropriate basis for government abstraction tariffs. The scope of abstraction charging includes water consumers abstracting directly for their own use, such as farmers, mining companies and manufacturers. One also includes water service companies that abstract not for their own use but in order to provide a public water supply to irrigators and to urban areas.

A useful analytic starting point is to establish that abstraction charges are a form of economic rent. The modern theory of rent was first developed in the early nineteenth century. Its application was to British agriculture and the common situation where capitalist farmers used land owned by the aristocracy and the gentry and paid rent for that right. At that time, agriculture contributed about one-third of Britain's total output. The classical school of economists argued that the appropriation of land has the consequent effect of the creation of rent. Thus, Ricardo writes: 'Rent is that portion of the produce of the earth, which is paid to the landlord for the use of the original and indestructible powers of the soil'.

It is true that we may now doubt that any power of the soil is indestructible. Nevertheless, since Ricardo's time, economists have used rent theory whenever they are dealing with a natural resource of economic value that can be privately appropriated and that is in restricted supply. Clearly, this makes rent theory applicable to natural supplies of freshwater. These flows are of economic value, rights in their abstraction can be privately appropriated and groundwater as well as surface water are in restricted supply in any given year.

However, the supposed inelasticity of the water supply deserves closer scrutiny. Within a catchment, short-term elasticity is perversely high where substantial stocks are held in a catchment's reservoirs or where aquifer stocks are high and groundwater pumping capacity is not fully utilized. But it is the elasticity of the planning supply function of water over the medium and long term that is at issue here. It is certainly true that total rainfall in the catchment less evaporation sets a hydrological constraint on long-term catchment abstraction, as Dubourg suggests. But the recycling and reuse of water weakens this natural barrier whilst the desalination of sea water and abstracted water imports from another catchment, can smash it. The problem is that the wider one casts the net for additional supply, the greater are its unit costs. It is the exponential character of this planning function which maintains the truth of the statement that, in the majority of the world's catchments with substantial populations, abstracted water is indeed in restricted supply.

Returning to the classical theory of differential rent, this assumed that competition existed between farmers for access to land and between landowners in supplying land. In such a situation, it is theorized that land rent on the least fertile tract of land worth cultivating would be zero and here the farmer would earn the going rate of profit on capital for the private sectors of the economy as a whole. Land which was more fertile (or better located) would bring the owner a higher flow of rental receipts such that the farmer, after paying this higher rent, would still receive only the going rate of profit. Rent could therefore be seen as the transformation of the surplus profits of capital, driven by the competition for land of differential fertility.

Ricardo had only a limited interest in non-competitive markets in the supply of land for rent. As a consequence, his theory is essentially demand-driven. This neglect of the supply-side determinants of land rent restricts the scope of differential rent theory in its application to abstraction charges levied by government. Such charges typically exist where an agency of the state licenses abstraction to specified individuals and institutions and where no property right exists to abstract ground and surface water without such a licence. Here, property rights on the supply side are assigned by a state monopoly. So analysis needs a means to interpret the legislative practice of charge-setting by state institutions in different catchments, regions or countries.

Having argued that abstraction charges are a form of economic rent, but that differential rent theory does not provide a basis to explain charge-setting by the state, the best way to proceed is to develop a taxonomy of charge-setting principles. Such principles can then provide the basis for interpretive empirical work in specific catchments, as well as a starting point for policy review of existing legislative practice. Various charge-setting principles are set out below and each is glossed in turn. Note that these charging principles are in respect of the right to appropriate the scarce resource for outstream purposes. They do not include the cost of supply-side infrastructure.

Nor are the public-sector costs of flood management, navigation, etc. included here:

- *No charge:* This is the lower limiting case. In the majority of the world's countries, government abstraction charges simply do not exist. In some cases the argument is that, since surface water and groundwater are a gift of Nature or of God, the state has no business in taxing it. This may explain the situation in Scotland, for example. In other cases, what is lacking is the administrative capacity to levy the tax.
- *A revenue-maximizing charge:* This is the upper limiting case. In principle government could raise all its revenue requirements from this single tax. In practice, of course, such a principle is never applied.
- *A market-clearing charge:* In countries and regions that are arid or semi-arid, or where levels of water consumption are high compared with effective rainfall, users may wish to gain access to more water than is available, at least in the absence of inter-catchment transfers. In

this case, government may put in place a demand-management policy in which a general abstraction charge is applied that, although it is not revenue-maximizing, does broadly match the demand by abstractors to the annual flows available. No pure examples exist of this, the closest being the public auction of tickets for fixed time and flow in the centuries-old water market of Alicante in Spain. Schiffler and his colleagues at the German Development Institute recorded for the case of Jordan that supply-fix policies in the early 1990s were under considerable pressure and that a demand-management philosophy was taking shape. Politically, it was impossible to tax water abstracted for use by farmers, but an abstraction tax on industry had been introduced there.

- *An environmental regulation charge:* In this case one is considering a country which has a well-developed national policy for water resource management. The necessity for a regulator of the freshwater environment is accepted and abstraction charges are levied and hypothecated to finance the costs of regulation. ('Hypothecation' is a term widely used in economics to refer to cases where government income from a defined tax is reserved for a specific expenditure category.) This charge may take the form of an average total cost levy equal to the total financial costs of regulation (including compensation payments for rescinded abstraction rights and any other miscellaneous items) divided by the total volume of water abstracted during the year.
- *A Pigovian charge:* Here, the principle is that government should set a charge which is differentiated according to the external costs imposed on society by each class of abstractor. K. William Kapp later formulated the parallel concept of 'social cost', covering 'all direct and indirect losses sustained by third persons or the general public as a result of unrestrained economic activities'. An example of the kinds of damage done by overabstraction is that of the Hueco Bolsón aquifer on the Mexican-USA border. Over a period of 70 years the water table fell by 25 metres (m), resulting in increased pumping costs, subsidence and contamination by increased flows of saline and polluted waters into the freshwater source. As a general rule, whilst abstraction charges may include a component in recognition of overabstraction, the *differentiation* of the charge on the basis of the monetary evaluation of the external costs never takes place because of the extreme difficulty of measuring them. Indeed, many institutional economists even deny that such measurement has any

meaning. Kraemer refers to the overwhelming problems of applying a Pigovian tax in his introduction to the development of German abstraction charging after 1988.

- *An incentives charge:* Incentive charging in a catchment or region can be defined as the use of a water tariff-a table of fixed charges-to give price signals to abstractors that reinforce water resource management based on environmental standards and regulatory controls. The best-developed system of incentive charging is probably that of the Federal Republic of Germany. An incentive tariff can include the following components:
  - A licence fee to have a new water abstraction installation approved by the regulatory authority. For example, Foster in their comprehensive introduction to the hydrogeological, legal and administrative aspects of groundwater licensing in Latin America indicate that installation fees are common and are hypothecated.
  - A charge per unit volume. This may be invariable with total volume consumed. Alternatively, it may rise with the volume drawn off. The total charge payable may be based either on the licensed volume or the actual volume abstracted. As Schiffler point out, a significant drawback of a rising block tariff is that it hits hardest those abstractors requiring large volumes simply because of the size of the farm or factory. It should also be observed that abstraction charging is never imposed where the installation's capacity falls below a minimum level set by government. This is the case at Keveral Farm.
  - A charge reduction for the quantity of water directly recycled to surface waters after use. To take the example of Didcot power station in the UK, in the mid-1990s this had a licence to abstract 142 million litres of water per day. The licence required 50-66 per cent of the water abstracted to be returned to the river, depending on flow conditions. Unit price was lower because of this non-consumptive use. In contrast, spray irrigation provided virtually no return flow to river or aquifer and so was undiscounted.
  - A charge that is greater for higher quality water. This is found in Germany for water drawn from deep aquifers.
  - A charge which varies with the seasons. The volumetric rate is higher in those months when demand is greater and higher too when precipitation is less.
  - A charge which is greater for certain locations. These include

(*a*) Upstream sources, because the length of the river exposed to abstraction impacts is greater,

(*b*) Rivers, lakes and aquifers most threatened by past or present overdraft;

(*c*) Regions with lower effective rainfall.

Abstraction Charges and Sustainable Catchment Management

The subject of abstraction charging is related to the debate on sustainability, beginning with the approach of Richard Dubourg whose contributions to the study of hydroeconomics are within the neoclassical framework. Dubourg suggests that aggregate sustainability is a situation where natural capital as well as non-natural capital are non-declining and specifically where 'critical capital' such as water, within the natural capital category, is non-declining.

From these definitions he deduces impeccably that aggregate sustainability is consistent with catchment management policies in which abstraction of surface water and groundwater is equal to effective rainfall. It has to be said that the adoption of such a definition of sustainable abstraction would be extraordinarily dangerous.

Abstraction at a rate equal to effective rainfall may be capable of leaving surface and groundwater stocks unchanged over the course of a year so that the critical capital stock is maintained. But a rate of abstraction at this maximum rate would have the effect of capturing the entire river flow at some point or points in the catchment. In effect, Dubourg's definition ignores completely what in the USA is termed the environmental need for water in river systems-all on the basis of three mathematical constraints. In *Introduction to the Economics of Water Resources* I have suggested an alternative, multi-dimensional approach. After defining the concept of a 'sustainable society', it is suggested that water resource planning in such a society has six principal fields of action:

- The protection of water's hydrocyclical capacity to renew its ground and surface-water flows and stocks;
- The conservation of society's species and natural habitats in all their fresh and saltwater environments;
- The husbandry of water in its supply and use;
- The supply of freshwater sufficient to meet the biological, cultural and economic needs of society's human populations;
- The purification of water from domestic, agricultural and industrial effluents;
- The drainage of water and the protection of rural and urban communities against flood.

Abstraction charges can contribute positively to the first three of these fields and, below, it is shown how this can be done, using what I shall call *full cost incentive charging*.

The preparation of a tariff scheme should take as its starting point the regulatory controls put in place by the catchment authority in carrying out its responsibilities. In the development of these controls, there is an important role for social cost-effectiveness analysis and social cost-benefit analysis. These techniques have a place, alongside environmental impact assessment, in comparing alternative regulatory options. The design of environmental regulation should be an economic process as well as an ecological one.

The tariff regime introduced to any specific catchment (or region) will be contingent on its physical geography, its habitats and species, its human settlement patterns, its economy and the power relationships which hold between various social and economic groups. So what we require are criteria for tariff design which can be applied in the appropriate way to any single area with all its unique characteristics. I propose three such criteria for full cost incentive charging.

- *First,* the annual income from abstraction charges should be hypothecated to the environmental regulator such that, when added to the income received from discharge fees, fishing licences, navigation permits, etc., abstraction fee income is sufficient to cover all the state's capital and current account expenditures on environmental regulation, research and database development, compensation payments, etc. In providing the regulator with hypothecated income, it is likely to increase their relative power as an agency of government. As Kraemer writes of Germany since 1988: 'On the whole, the water resource taxes contributed to capacity building within the water management administration in the German Länder and thus partly overcame the implementation deficit in water resource management'. It will also provide a budget to finance the legal costs of modifying or terminating abstraction licences that threaten the hydrocycle or undermine nature conservation. Higher abstraction charges will also give a price signal to water companies to reduce their storage and distribution leakage between the points of abstraction and the user's gate.
- *Second,* where the full cost abstraction tariff still leaves an excess of demand for abstracted water over its licensed supply, the charge should be raised so that market clearing takes place.
- *Third,* specification of the components of the abstraction charge should provide incentives for abstraction behaviour that is economically efficient and that avoids environmental degradation. Price per unit

volume should be invariant with total volume abstracted, unless there are countervailing economic or environmental arguments. Price should be discounted where abstractors recycle their off-take to surface or groundwater sources. (Discharge fees should be used to handle the water quality aspects of recycled water.) Charges should be higher for abstracted water of higher quality. Seasonal variations in effective rainfall and economic demand should be dealt with through the licensed volume provisions laid down by the regulator and by the market-clearing criterion. Charges should be higher for upstream sources and for abstraction in locations where species and habitats are more threatened by abstraction.

The institutional framework that would be most appropriate for full cost incentive charging deserves discussion and my ideas here have been strongly influenced by Karin Kemper's *The Cost of Free Water*. The basic approach is a negotiation model in which there exists a public-sector catchment agency that, through a negotiating forum, develops its policies with the advice of water companies, direct abstractors, environmental organizations and water user associations representing the domestic sector, irrigated agriculture, mining and manufacturing, etc. The original prototype for such negotiation models is the French water parliaments. Central government retains the statutory right to determine which public and private institutions may enjoy the right to abstract water, on what scale, in what locations, at what time of year, at what price-but it delegates such rights to the catchment agency.

The agency, with the assistance of its partners in the negotiating forum and with a full understanding of existing customary rights, then assigns formal abstraction rights on a time-limited basis to specific abstractors or groups of abstractors. The time limit would be 10 years, rolled over each year except where the agency allows the licence to expire. Where these 10-year rights need to be modified or rescinded for hydrological, environmental or economic reasons, compensation would be payable to the abstractor so affected. Rules would also exist to address third party impacts. Abstraction rights assigned to abstractors could be freely traded provided the agency had approved such transfers in the light of their social, economic and environmental impacts. Full cost incentive charging would be the basis of the price paid by abstractors for their water.

It has the objectives of:

- Underpinning environmental regulation with a hypothecated income source;
- Requiring abstractors (and therefore final users) to cover the full costs of regulation;

- Bringing the quantity of water demanded by abstractors into line with regulatory limits; and
- Giving price signals that promote both allocative efficiency as well as abstraction practices that avoid damage to riverine eco-systems.

No charge would be made for abstractions below a minimum scale-the transaction costs of such charges would be high compared with the volumes abstracted.

Arrangements would be made to monitor abstraction with respect to its location, time and quantity, as well as to invoice abstractors, to collect the charges owed and to enforce all agreements. Such transaction costs would be included in the full costing tariff of abstraction charges. The creation of this institutional framework imposes social, economic and political costs on the parties concerned, structural costs of change both real and financial in their nature. Therefore the negotiating forum may agree that it is sensible for full cost incentive charging to be phased in gradually.

Irrigation Management Transfer on Mexico

Local associations of water users serve the same functions as irrigation agencies, but on a very localized level. In some countries such as Nepal and the Philippines, the major portion of the irrigated area is managed locally, through village-based water user associations. Typically such associations manage very small irrigation canals that were constructed by the users, perhaps centuries ago and the associations have grown up around the need for operating and maintaining these canals. In many developing countries such traditional canal systems have been the target of modernisation efforts by government agencies, often funded through international development assistance. The relatively crude physical infrastructure of many of these canal systems was rebuilt and absorbed into the domain of the government agency, which replaced the management functions of the indigenous water user associations.

Today, local water user associations are recognised as resources of 'social capital' which will have an increasingly important role to play in the coming decades. Instead of absorbing such associations into the state organisation, the capacity of associations is developed so that they can improve the performance of their own irrigation and drainage systems. This participatory approach was re-invented in the Philippines in the 1970s and has since evolved into a global trend that combines participation with various degrees of privatisation. In many developing countries a major focus of this process is the institutional challenge of establishing new water user associations which can serve the management functions previously handled by government.

A particularly fruitful source for a case study of this trend is Mexico. With 5 million irrigated hectares it was the seventh largest irrigator in the world in 1995. It began the irrigation management transfer process as early as 1989 and the experience has been extensively researched. Finally, Mexico has for a number of years been considered as a paradigm for this form of institutional transformation. The case study draws heavily on the work of Kloezen, Garcés-Restrepo and Johnson.

The broad political and economic context within which Mexico introduced irrigation management transfer was central government's neoliberal response to the economic crisis of the 1980s and the unsatisfactory performance of the irrigation districts under the management of the National Water Commission. Transfer has been a top-down process motivated by the international development banks, the main objective being to reduce public expenditure on irrigation OM&M whilst promoting greater user participation in irrigation management. It has been accompanied by a new National Water Act, revision of the Constitution to give legal foundation for the privatization of the *ejidos* (land reform communities) in all irrigation districts and termination of guaranteed crop prices and subsidized credit. 'Dismantling the public sector, including the public irrigation sector, would not have been possible without the commitment at the highest political levels to reduce staff working in the public sector'.

The institutional transformation has been rapid and radical. In its first phase the National Water Commission retained management of the headworks and the main canals. Water user associations took over financial and managerial responsibility for OM&M below the main canal. In the second phase the Commission is left with management of the reservoirs and surface-water-pumping stations; a District Federation of the user associations takes on the OM&M of the main canals. These associations are non-governmental local entities for collective action.

The high speed of the process and the low resistance of farmers have surprised many observers. It seems that transfer was preceded by other neoliberal reforms in agriculture. Farmers knew the Commission's traditional OM&M was to be terminated and OM&M had in any case declined in quality. Reform came first in the larger districts in the north, where many large private producers were known to support change. Lastly, the programme built on a strong organizational base-the *ejidos* themselves and the private farmers' co-operatives and unions.

Government informed farmers of the impending change-there seems to have been no serious consultation-and instructed them to select their delegates to the new user associations. Thereafter, delegates to each association elected from among themselves a president, a treasurer and a secretary. The Commission worked with the new user groups for 6 months or more from the time of transfer and provided their leaders and technical staff with extensive training in OM&M and financial management. The Commission also granted concessions to the associations to use its machinery and equipment so that there would be no capital expenditure required in advance.

Prior to transfer the National Water Commission was wholly responsible at the district level for the planning of annual and seasonal water allocations. Similarly the Commission employed the heads of the irrigation units into which each district was divided. These units were more or less independent hydraulic blocks with sizes ranging from 3,000 to 20,000 ha. Unit heads and their channel-keepers were responsible for daily OM&M at all system levels down to the farm inlets. Farmers paid their fees at the unit office and were given water by the channel-keeper.

After transfer, hydraulic committees were introduced at district level for allocation planning; these committees were composed of representatives from the Commission, the state government and each user association in that district. At the same time the district 'units' were replaced by 'modules'-two or more per unit-and it is at this level that the new water user associations manage.

They collect irrigation water fees directly from farmers and employ their own module managers, channel-keepers, maintenance personnel and administrative staff. Based on the volume of water a module takes from the Commission and on the proportional amount of main infrastructure serving a module, each association must pay a percentage of the total fees it collects to the Commission.

The specific subject of this case study is the Alto Río Lerma Irrigation District. Located in the State of Guanajuato, it has an area of 113,000 ha. There are some 24,000 farmers of whom 55 per cent are from the *ejidos* and 45 per cent are private growers. The average landholding is 5 ha. Private holdings average twice the size of *ejido* holdings. The district's climate is sub-humid with annual precipitation of 750mm and potential evapotranspiration of 1900mm. Eighty millimetres of rain falls in the winter season from November to April and 670 mm falls from May to November. Average temperature is 19 °C and relative humidity is 60 per cent.

The Alto Río Lerma Irrigation District enjoys access to both surface water and groundwater. The river has four storage dams with a combined capacity of 2,140 mcm, as well as five diversion dams. The distribution network has 475 km of main canals and 1,660 km of secondary and tertiary canals, as well as 1,030 km of drainage canals. There are 1,710 deep wells exploiting three different aquifers with a total annual recharge of 500 mcm. Seventy per cent of the irrigated hectarage uses surface water and 30 per cent uses groundwater. Wheat and barley are the main crops in the dry winter months. Sorghum, maize and beans predominate in the wet summer season. All farmers grow vegetables and the private growers do so for the export market.

At the start of each agricultural year in November the hydraulic committee decides on the total area that can be safely irrigated in the district and by each of the eleven modules. This total area is derived from the combined volume of water in the four storage dams. The committee next sets the number of times that irrigation services can be delivered to each farmer in each season, typically 3-5 times in the winter and once in the summer. Each module receives its allocated share of surface water over the year and, given the module's ability to restrict the area irrigated by users, farmers can request water at any time within the constraints of the seasonal maximum. Farmers pay a fee to their water user association prior to each individual irrigation and get a receipt that has to be shown to the channel-keeper before water is allocated to their fields.

Kloezen and his colleagues show a keen awareness of the political and macroeconomic context of irrigation management transfer as well as a familiarity, for example, with Fox's sociological work on building social capital from below. However, the language of common property resources, freeriding and transaction costs is absent from their reports. What we do learn is that the number of staff engaged in governance, OM&M, administration, monitoring and evaluation *rose* by 13 per cent between the 'before' and 'after' transfer periods of 1992 and 1996. The proportion of persons employed by the National Water Commission dropped from 100 per cent to 38 per cent. The associations feel that the Commission staff numbers are excessive and that:

...an unspecified percentage of these...staff are residual personnel that for political and labour-union-related reasons remain within the agency with no specific task. This has been one of the major reasons why the modules wanted to create the [District Federation of the user associations] which will take over management of the main system. They feel that this will be much more cost-effective than the current arrangement.

As already indicated, the water user associations hire all their own staff. In doing so they have shown reluctance to employ ex-Commission personnel. The associations argue that Commission channel-keepers and other staff were often out of control, poor workers, unaccountable, given to rent-seeking behaviour and were likely to involve the trade unions in module management in a divisive manner.

To what degree do farmers accept that the shared use of the resource is now equitable? The importance of the new, district-wide, hydraulic committees should be noted, but the allocation planning method has not changed at this level. However, there is evidence that correspondence between the volume of water assigned to modules and the actual volumes received by them is better than before transfer. This is probably because every association was represented on the hydraulic committee, with each module insisting on receiving the volume it had been assigned and had paid for.

With respect to individual farmers, a detailed, post-transfer study shows that in the sample area 'there is no clear bias towards head- or tail-end farmers and that all farmers receive sufficient water to meet crop requirements'. In a sample of farmers, 34 per cent said water distribution amongst them was poor before transfer and good after it, compared with only 15 per cent who held the reverse opinion.

Probably the most important contribution to equitable distribution of the resource has come from the associations' control over channel-keepers. Bribes exacted by Commission employees under the previous regime were impossible for farmers to eradicate. The downward shift in power to water users now gives them the control over rent-seeking behaviour that they lacked.

To what degree has infrastructural operation and maintenance been improved? With respect to operation, the post-transfer study indicated an improvement in water adequacy at field level, in the timeliness of water delivery and in access to the channel-keepers. In the sample of farmers, 40 per cent said the service provided by the channel-keepers was poor before transfer and good after it, compared with only 14 per cent who held the reverse opinion. Eighty-three per cent of farmers believe that the water user associations should now retain responsibility for operation of the main and secondary systems.

The transfer process included free concessionary use of Commission machinery such as draglines and hydraulic excavators. This was a flying start for users. Significantly, some of this equipment was in serious disrepair so the modules bought twenty-nine pieces of new heavy machinery from their user fee income and also received equipment from a World Bank programme.

Maintenance expenditure in constant peso prices almost doubled in the before- and after-transfer comparison. There was an approximately threefold increase in the desilting of secondary canals and drains. Farmers' perceptions of the condition of the irrigation and drainage network were that the network had improved considerably.

To what degree have negative externalities arising from individual farmers' actions declined since the user associations were set up? Kloezen and his colleagues address this question only with respect to overabstraction of groundwater. By customary practice, modules exclude from the surface-water supply those areas that have access to wells. But groundwater users are permitted to use the canal network to distribute their pumped supplies.

Unfortunately the aquifers in the district are definitely being overexploited by about 20 per cent of annual recharge. By 1996 the water table was falling by 2-5 m annually. Groundwater mining does not appear to have become worse since transfer, but transfer has been unable to diminish it.

To conclude this case study, the new arrangements for the financing of capital and current costs are examined. We have already seen that the association pays fees to the Commission for its services, that the association charges the farmer for each irrigation, that there has been a decline in bribes paid to channel-keepers and that charges to farmers have been used not only for prime costs but also for equipment purchase.

Forty per cent of farmers surveyed report that payment procedures are now less cumbersome compared with only 2 per cent holding the reverse opinion. Sixty-nine per cent stated that bribery of channel-keepers had been reduced. But many farmers still come to the module office to complain of keepers demanding private payments and this has often led to workers being dismissed. Keepers are also rotated between module sections to prevent the growth of patron-client relations. Nevertheless as many as 30 per cent of farmers surveyed were willing to admit that they bribe keepers to irrigate more than the farmer's entitled area or to get water at different times from those programmed.

So the indirect charges are higher than the official ones-as are the wages of keepers. Under the Commission, charges were way below prime costs, let alone the full cost of the service. To prepare the way for a more self-reliant organization, the Commission increased the charge by some 400 per cent 2 years prior to transfer. It then remained unchanged in nominal pesos through and after transfer. But inflation halved the real value of the charge in the first 2 years of the associations' existence.

What is undoubtedly impressive is that in terms of the ratio of fees actually collected to actual OM&M expenditure, both given in pesos at current prices, the rate jumped from an average of 50 per cent in 1989-91 to 123 per cent in 1993-6. These facts remind us that in irrigation economics the estimation of the cost of water to farming families and agribusinesses should be based on the payments *actually* made, not on those that are *supposed* to be made. Kloezen and his colleagues never attempt to measure what proportion of the full cost of irrigation services is met by farmers' actual payments, but it seems likely to be considerably less than 100 per cent. This raises the question: 'What organisation in future will be responsible for infrastructural investment and rehabilitation, how will it raise the required finance and how will these costs be repaid?'

For the 1996-7 winter season the associations used the hydraulic committee to raise the irrigation fee by more than one-third. They also pushed the Commission to transfer OM&M responsibility for the main canal to a newly created District Federation of the user associations. In 1997 the average percentage of farmers' irrigation fees paid to the Commission was correspondingly slashed from 25 per cent to 9.5 per cent.

Tradable Abstraction Rights in Australia

Up to this point, consideration of abstraction has covered the right to abstract and the charges that a catchment agency might levy on farmers for the enjoyment of that right. *Tradeable* abstraction rights (TARs) have been discussed only briefly-although long enough to recommend that, with appropriate safeguards, such rights should be widely introduced. This case study reviews the introduction since the early 1980s of TARs within irrigated agriculture by the State of Victoria in Australia. The case study updates by 4 years one that I published in 1997.

In every catchment, some users of water will value their abstraction rights more than other users, for example because they employ water more productively. In that situation, trading in such rights can be financially attractive to both the existing owner of such rights and the party that wishes to see them transferred. To understand trading in abstraction rights we need to come to grips with the relevant law, with differential valuations of water and with the legal, hydrological and engineering means for implementing transfers. In this field, valuable work has been carried out by Robert Hearne and William Easter, in Chile, although when they were writing their report less than 1 per cent of all abstractions operated through such water markets there.

In Australia, the practice is now extensively developed and there the main objective of TARs has been the more efficient use of scarce irrigation water. The Australian water economy is said to be in its mature phase, where the long-term average total cost function is rising sharply, where there is intense competition for existing supplies and where the hydrosocial infrastructure requires costly rehabilitation.

Victoria is one of the constituent states of federal Australia and lies in its south-east corner. The dominant physiographic feature is the Great Dividing Range, running east-west in the eastern two-thirds of the state.

This creates a rainshadow in northern Victoria, where average annual rainfall is 450 mm in the Goulburn-Murray Irrigation District (GMID). This is the largest irrigation area in Australia, covering some 820,000 ha and it represents the bulk of water trading in the state. Irrigation is sourced from reservoirs on regulated surfacewater flows. Meat and dairy products are the principal agricultural output; higher-valued crops such as horticultural products are constrained by the red-brown soil type. Surface-water resources have been extensively developed in the past and any further supply growth would require inter-basin transfers. The GMID is principally designed to provide security of supply during a prolonged series of drought years. The main water dams are operated on a carry-over basis, where water is accumulated in years of high river flows for use in drier years.

Australia's modern history of property rights in water begins in the 1880s, when Victoria introduced new water laws based on the recommendations of a Royal Commission chaired by Alfred Deakin. He proposed that water allocations should be tied to the land, that rights to water should be vested in the British Crown and that allocations to landholders should be the responsibility of the state governments. Riparian owners retained limited common-law rights for domestic use, stock watering, gardens and a maximum of 2 ha of irrigated land for fodder crops. Over the next 15 years, similar legislation was introduced in the other federal states. Using the new legislation, the states ventured into large irrigation projects, providing water to farmers far below average total cost. Areas were established in which government constructed massive irrigation and drainage infrastructures, such as the Goulburn-Murray Irrigation District itself, the Murrumbidgee Irrigation Area in New South Wales and Riverland in South Australia.

The subsidy of water led to overuse and this was consolidated by the historical overallocation of water to irrigated farms. As argued in the work of Pigram and co-workers: This has its roots in the social objectives of past governments for the development and use of water resources. The overriding objectives were an equitable distribution of water among farms and the promotion of regional development and closer settlement of inland areas. All farmers were considered to have an equal right to the available water, irrespective of how much was needed to irrigate the proposed crops to be grown on the farms, or of any consideration of how efficiently the water would be used. As water rights became capitalized into land values, individuals had the economic incentive to retain their entire water rights through demonstrating a history of use. In many cases, this actually translated into a history of over-use.

By 1990, the arrangements in the GMID were that irrigators paid for a water right based on both the amount of land they held that was suitable for gravity-fed irrigation and the irrigation district in which they were located. An annual charge applied to the water right, invariant with the volume of use. Subject to availability, 'sales water' was also available, at a volumetric charge based on the water right charge. These water rights in their specific form were assigned by a licensing system for a fixed period of 15 years, with an expectation of reissue. The authorities could vary the volume licensed, usually in times of shortage. However, before 1987-8, no arrangements existed for *transferring* farmers' abstraction rights.

An Australian interest in TARs (also known as tradeable water rights or transferable water entitlements) began to surface in the mid-1970s. For some, the introduction of TARs seemed to offer a new flexibility to existing arrangements, with clear economic and environmental advantages. *First,* we have already seen that the maturing of the Australian water economy in recent decades is associated with high average total cost for the long-term supply curve. So, TARs offered a new direction that, by reallocating water supply, would reduce the pressure for aggregate supply expansion. This reallocative effect would not be merely within the farming sector. There was also the prospect of reducing agricultural overuse to open up supplies for a range of urban and industrial uses.

Second, supply reallocation within agriculture was a major objective. There was a widespread belief by the mid-1980s that TARs would switch water from lower to higher water-productivity uses in the farming sector. In Australia, at that time, the agricultural sector accounted for 80 per cent of total water use. In Victoria, each year, up to one-third of irrigators were using less than their full water right allotment. Specific switches into river red gum watering, salinity dilution and dairy farming were forecast. *Third,* environmental benefits were expected from transferability. The new policy promised to reduce the scale of infrastructure construction for inter-basin transfers, with all their negative externalities. Development proposals were meeting increased resistance from environmental groups. Moreover, reductions in agricultural overuse promised to have a positive environmental effect through a reduction in waterlogging, salinization and biocide dispersion.

The redistribution of abstraction rights could have been sought by the administrative processes of the licensing system. But this would have been met by political resistance from the farms on which volumes were to be reduced and land values consequentially cut. Tradeable abstraction rights offered these farms a pay-off.

For landholders wishing to move out of irrigation, but to remain in dryland agriculture, transferability allows water entitlements to be sold separately from the land. Previously, the options for such irrigators were either to cancel their water rights licence (or not renew it) and get nothing for the right, or sell the entire irrigation holding and buy a dryland property elsewhere. Transferability can permit a more flexible retirement plan for an irrigator, or facilitate a long-term change in enterprise or financial structure of the farm business.

The fundamental requirement of a workable and efficient market in abstraction rights is a clear specification of property rights in water such that:

- Rights in land and rights in water must be separable;
- The volume of water that an individual or institution has available for transfer must be clearly stated, as well as any special conditions on its use-such volumes are likely to be conditional on rainfall, or surface flows or groundwater stocks;
- The right to transfer such water at a privately negotiated price must exist;
- The period over which such a transfer is deemed to be effective, temporary or permanent, must be known;
- The power of government to restrict or terminate abstraction rights at a future date must be known.

In Victoria, TARs were cautiously introduced in 1987-8 with a *temporary* scheme, before permanent transfers were considered. In the GMID the transfer period was to be for 1 year, only between irrigators and only within the same supply system. There was no volumetric limitation but stock and domestic allocations had to be retained. A state agency assessed possible third-party effects and could refuse the transfer if these were significant. The arrangements were not to affect significantly delivery and drainage channel capacity or salinity. Finally, the price was determined between buyer and seller but the agency's fixed fee was A$70.

A new Water Act in 1989 permitted *permanent* transfers of abstraction rights between farmers, with effect from 1991-2. Transfers out of agriculture were still proscribed. However, despite these references to 'permanence' in trading, the state assignation of water rights through the licensing system, described above, meant that a purchaser of abstraction rights could not be assured in law that such water rights would continue indefinitely. The state government retained long-term flexibility in its management powers.

TARs impose certain transaction costs. In the GMID case they seem to be modest because of the legitimacy of the institutions, the support from farmers for the general approach to the new rights and the competence and honesty of Victoria's state administration. Incremental supply costs also exist if distribution and drainage structures cannot handle the increased water volumes. However, in Australia, water agencies have taken the easy administrative option of simply refusing transfers where existing capacities would be exceeded.

The outcomes of Victorian TARs in the early years after they were first introduced are considered next. The scale of change can be measured by the volume of abstraction transfers as a percentage of all abstractions. With respect to benefits, we are looking for: the avoidance of new infrastructure expenditure as a result of the redistribution of abstractions within and between user categories; an increase in agricultural productivity as a result of abstraction redistribution within farming; and a fall in waterlogging, salinity and pollution from run-off because of the reduction in agricultural overuse.

In the 2 years 1987-8 and 1988-9 the trading of abstraction rights in Victoria (most of it in the GMID) averaged 25 mcm, less than 2 per cent of total abstractions. Price per megalitre ranged from A$8 to A$20. Gross margins increased. But at this point the judgement was that 'Despite widespread endorsement of the concept in Victoria, transfer activity has been sporadic in that state'.

In Victoria by 1990-1 trade was still 'insignificant compared to the total volume'. In a mail survey of all 299 permanent water transfers within the GMID up to 1994, with a 63 per cent response rate from buyers, purchases made up between 0 per cent and 2.3 per cent of total allocations when the responses were grouped by district. What the survey showed was that sellers were releasing sleeper water (*i.e.* unused allocations) and that farmers in financial distress were also selling. In a smaller number of cases, farmers either wished to cut their irrigation agriculture, or to farm without irrigation, or to retire. In respect of purchasers, the single most important reason for buying water, applicable to 65 per cent of respondents, was that they 'wanted to secure existing crops against future drought'.

Stringer suggested that in Victoria as a whole in 1991-4, the average percentage of allocated water traded in each of those years was 0.33 per cent. In the Murray region as a whole in 1993-4, the volume of transfers was 17 mcm for temporary transfers and 8 mcm for permanent transfers. For 1994-5, these figures changed dramatically.

Temporary transfers soared to 265 mcm whereas permanent transfers fell to 2 mcm. At this point it could be argued that by the mid-1990s TARs in Victoria had created a space for single-season switching of abstraction, most active in years of low rainfall. By the year 2000 Australian statistical reporting of TARs had become much fuller and more informative as a result of work by the Australian National Commission for Irrigation and Drainage. Its *Benchmarking Report* contains data for nine Goulburn-Murray systems known as Murray Valley, Shepparton, Central Goulburn, Rochester, Pyramid-Boort, Torrumbarry, Nyah, Tresco and Woorinen. In all of these except the last the percentage of surface-water irrigation supply points that are metered lies in the range 87-98 per cent.

A broking service exists in all nine systems to facilitate the trading process between users. In 1998-9 temporary trading took place in all nine systems, averaging 11 per cent of total abstraction entitlements for that year. Six of the nine systems reported permanent trading that averaged 2 per cent of entitlements in those six areas.

From the perspective of this book, the material on why one set of irrigators sell their TARs and others purchase them is of particular interest. A substantial paper by Björnlund and McKay throws some light here.

The authors state that the main use of irrigation water is in the cultivation of pastures for dairy cattle. Some activity is found too in horticulture and viticulture. Mixed grazing and cropping is said to have low-value water use. Purchasers of tradeable abstraction rights were shown to be seeking to ease their reliance on annual sales water and temporary water market purchases. TAR sellers are smaller than buyers in terms of irrigated area; water trading can be seen to be integrated with a long-term consolidation and amalgamation of farm properties.

Sellers were asked how they used the cash proceeds of their permanent sales-note that a permanent sale does not imply the sale of a farm's entire water entitlement:10 per cent was invested in laser grading, 5 per cent in improved drainage and another 5 per cent in other irrigation equipment. Fifty per cent of the sample said the money was used for general consumption purposes, related or not to the farm as an enterprise; 21 per cent went to debt reduction; 6 per cent to non-irrigation equipment; and 3 per cent for the purchase of another farm.

Clearly the subject is ideally suited to an irrigation economics case study, for both the permanent and temporary markets, into the differential motivation of buyer-seller *pairs*.

A Blocked Channel

One aspect of this blocked channel concerns payments for abstraction rights. I have given a very full account of the criteria that can be used by a state authority in constructing an abstraction tariff for farmers and the relevance of such a tariff to effective catchment management. But there is no accompanying case study to test these ideas-merely brief, illustrative examples. In contrast there is an extended discussion of the recent history of tradeable abstraction rights in Victoria, but the theoretical material is adhoc and fragmented.

Despite this the research reports of Wim Kloezen, Carlos Garcés Restrepo and Sam Johnson seem to me to be wholly successful in coming to grips with the dynamic institutional changes of this catchment. Perhaps it is because they were a team with varied disciplinary backgrounds, which is so important in water resource planning. Perhaps it is that irrigation management transfer is quintessentially a compound of private economic interests and political power set within a given hydrological context-and that Kloezen and his colleagues have a good grasp of hydropolitics. In contrast, the language of many economists is so distant from the complex drives of individuals, families and organizations and so wedded to a vision of the determining interplay of measurable quantities, that they fail to understand the nature and exercise of power.

SOURCES OF IRRIGATION WATER

Sources of irrigation water can be groundwater extracted from springs or by using wells, surface water withdrawn from rivers, lakes or reservoirs or non-conventional sources like treated wastewater, desalinated water or drainage water. A special form of irrigation using surface water is spate irrigation, also called floodwater harvesting.

In case of a flood (spate) water is diverted to normally dry river beds (wadis) using a network of dams, gates and channels and spread over large areas. The moisture stored in the soil will be used thereafter to grow crops. Spate irrigation areas are in particular located in semi-arid or arid, mountainous regions. While floodwater harvesting belongs to the accepted irrigation methods, rainwater harvesting is usually not considered as a form of irrigation. Rainwater harvesting is the collection of run-off water from roofs or unused land and the concentration of this. Some of Ancient India's water systems were pulled by oxen.

Around 90% of wastewater produced globally remains untreated, causing widespread water pollution, especially in low-income countries. Increasingly, agriculture is using untreated wastewater as a source of irrigation water. Cities provide lucrative markets for fresh produce, so are attractive to farmers.

However, because agriculture has to compete for increasingly scarce water resources with industry and municipal users (see Water scarcity below), there is often no alternative for farmers but to use water polluted with urban waste, including sewage, directly to water their crops. There can be significant health hazards related to using water loaded with pathogens in this way, especially if people eat raw vegetables that have been irrigated with the polluted water. The International Water Management Institute has worked in India, Pakistan, Vietnam, Ghana, Ethiopia, Mexico and other countries on various projects aimed at assessing and reducing risks of wastewater irrigation. They advocate a 'multiple-barrier' approach to wastewater use, where farmers are encouraged to adopt various risk-reducing behaviours. These include ceasing irrigation a few days before harvesting to allow pathogens to die off in the sunlight, applying water carefully so it does not contaminate leaves likely to be eaten raw, cleaning vegetables with disinfectant or allowing fecal sludge used in farming to dry before being used as a human manure. The World Health Organization has developed guidelines for safe water use.

Water Scarcity

Fifty years ago, the common perception was that water was an infinite resource. At that time, there were fewer than half the current number of people on the planet. People were not as wealthy as today, consumed fewer calories and ate less meat, so less water was needed to produce their food. They required a third of the volume of water we presently take from rivers. Today, the competition for water resources is much more intense. This is because there are now more than seven billion people on the planet, their consumption of water-thirsty meat and vegetables is rising, and there is increasing competition for water from industry, urbanisation and biofuel crops. To avoid a global water crisis, farmers will have to strive to increase productivity to meet growing demands for food, while industry and cities find ways to use water more efficiently.

Successful agriculture is dependent upon farmers having sufficient access to water. However, water scarcity is already a critical constraint to farming in many parts of the world. With regards to agriculture, the World Bank targets food production and water management as an increasingly global issue that is fostering a growing debate. Physical water scarcity is where there is not enough water to meet all demands, including that needed for ecosystems to function effectively. Arid regions frequently suffer from physical water scarcity. It also occurs where water seems abundant but where resources are over-committed. This can happen where there is overdevelopment of hydraulic infrastructure, usually for irrigation. Symptoms of physical water scarcity include environmental degradation and declining groundwater. Economic scarcity, meanwhile, is caused by a lack of investment in water or insufficient human capacity to satisfy the demand for water. Symptoms of economic water scarcity include a lack of infrastructure, with people often having to fetch water from rivers for domestic and agricultural uses. Some 2.8 billion people currently live in water-scarce areas.

How an in-ground Irrigation System Works

Most commercial and residential irrigation systems are "in ground" systems, which means that everything is buried in the ground. With the pipes, sprinklers, emitters (drippers), and irrigation valves being hidden, it makes for a cleaner, more presentable landscape without garden hoses or other items having to be moved around manually. This does, however, create some drawbacks in the maintenance of a completely buried system.

Controllers, Zones, and Valves

Most irrigation systems are divided into zones. A zone is a single irrigation valve and one or a group of drippers or sprinklers that are connected by pipes or tubes. Irrigation systems are divided into zones because there is usually not enough pressure and available flow to run sprinklers for an entire yard or sports field at once.

Each zone has a solenoid valve on it that is controlled via wire by an irrigation controller.

The irrigation controller is either a mechanical (now the "dinosaur" type) or electrical device that signals a zone to turn on at a specific time and keeps it on for a specified amount of time.

"Smart Controller" is a recent term used to describe a controller that is capable of adjusting the watering time by itself in response to current environmental conditions. The smart controller determines current conditions by means of historic weather data for the local area, a soil moisture sensors (water potential or water content), rain sensor, or in more sophisticated systems satellite feed weather station, or a combination of these.

Emitters and Sprinklers

When a zone comes on, the water flows through the lateral lines and ultimately ends up at the irrigation emitter (drip) or sprinkler heads. Many sprinklers have pipe thread inlets on the bottom of them which allows a fitting and the pipe to be attached to them.

The sprinklers are usually installed with the top of the head flush with the ground surface. When the water is pressurized, the head will pop up out of the ground and water the desired area until the valve closes and shuts off that zone. Once there is no more water pressure in the lateral line, the sprinkler head will retract back into the ground. Emitters are generally laid on the soil surface or buried a few inches to reduce evaporation losses.

Problems in Irrigation

Irrigation can lead to a number of problems:

- Competition for surface water rights.
- Depletion of underground aquifers.
- Ground subsidence (*e.g.* New Orleans, Louisiana)
- Underirrigation or irrigation giving only just enough water for the plant (*e.g.* in drip line irrigation) gives poor soil salinity control which leads to increased soil salinity with consequent build up of toxic salts on soil surface in areas with high evaporation. This requires either leaching to remove these salts and a method of drainage to carry the salts away. When using drip lines, the leaching is best done regularly at certain intervals (with only a slight excess of water), so that the salt is flushed back under the plant's roots.
- Overirrigation because of poor distribution uniformity or management wastes water, chemicals, and may lead to water pollution.
- Deep drainage (from over-irrigation) may result in rising water tables which in some instances will lead to problems of irrigation salinity requiring watertable control by some form of subsurface land drainage.
- Irrigation with saline or high-sodium water may damage soil structure owing to the formation of alkaline soil

7

Strategic Planning and Management of Water Resources

INTRODUCTION

Lessons from the Application of SPM to Water Resources

The key results and outcomes of the project "Capacity-building in Strategic Planning and Management of Natural Resources in Asia and the Pacific" in the water sector can be summarized in the following three aspects: important achievements of the project, key obstacles in application of SPM, and important lessons learned and best practices.

The important achievements of the project can be described in three main aspects: tool development, interest generation, and practical application of the methodology.

Tool Development

The development of the Guidelines marked an important step in the application of strategic planning and management approach to water resource, since it was the first time that a specific set of guidelines were specifically developed for the water sector and derived mainly from the practical experiences of the region. This fact can be illustrated by the translation of the Guidelines into Chinese, Russian and Vietnamese by the respective technical agencies. In addition, several enquiries were addressed on the guidelines on application to water resources management in the forestry sector in Malaysia and to the development of integrated river basin management modules for the UNESCO-IHE Institute for Water Education.

Generated Interest in SPM

The interest generated by the project on SPM of water resources was higher than expected in terms of the number of requests for case studies, which was higher than the project could deliver, although the actual number of case studies was much higher than the original number of case studies planned for the project, 17 in comparison with 10. Of particular interest was the decision of the Vice Minister of Water Resources of Uzbekistan to immediately apply the concept at the subregional workshop.

Best Practices of SPM

The seventeen case studies on the application of SPM to the water resources sector were conducted for the practical purpose of formulation of related policies and strategies at all the levels as recommended at the regional and subregional workshops: national (Myanmar, Thailand, Sri Lanka, Mongolia), basin (China, Fiji, Malaysia, Papua New Guinea, Viet Nam and five countries in Central Asia), sectoral level (Pakistan) and organizational (the Lao People's Democratic Republic and Philippines). Out of these case studies, the momentum generated by the application for the case studies in Fiji, Myanmar and Viet Nam proved the immediate impacts of the project, in terms of immediate actions taken by the agencies concerned.

EXAMPLES OF GOOD PRACTICES

The gross domestic product (GDP) of China increased substantially in the early 1990s, with industry and manufacturing contributing a large proportion of this growth. The rate of GDP growth was recorded at 6.5 per cent from 1960 to 1970, 9.7 per cent between 1977 and 1985, and 8.7 per cent in 1996 and 1997. The growth rate in foreign trade was equally impressive during the same periods at 14.3 per cent, 15.2 per cent and 7.6 per cent, respectively.

In the Five-Year Plan (2000-2005), the Government of China has set the objectives of developing and maintaining a high quality of life for all. In the process of developing the management of its strategic plan, the Government has specified the following five goals:

- Achieving an average economic growth rate of 7 per cent annually, which will translate into a per capita GDP of RMB 9,400 by 2005;
- Optimizing the industrial infrastructure, and strengthening international competitive capacity;
- Increasing research and development expenditure to more than 1.5 per cent of GDP, and improving education;
- Controlling population growth to keep it within 1.33 billion by 2005, improving the eco-environment and conserving resources; and
- Achieving an average annual per capita income of about 5 per cent, and improving health and social welfare.

These development goals clearly require support from the water sector, which comprises a number of agencies with resource development and management functions related to water resources. The sector is largely structured around the various uses of water, such as hydropower, irrigation, water supply, fisheries and wastewater. These water subsectors have developed their programmes with the aim of providing the support required for China's development objectives. These programmes, however, do not consider the potential for interaction between the policies and plans of the subsectors, because no agency was given overall responsibility for water resources management or coordination of water development.

China Water Sector

China is located in south-eastern Eurasia and covers an area of 9.6 million km^2. Affected by ocean-land distribution and topographic conditions, China is characterized by a monsoon climate. South-eastern China is pluvial while the north-western region is dry with the main period of rainfall being concentrated in the summer season. China has a large number of rivers, totalling some 50,000 with catchments larger than 100 km^2 and 1,500 rivers with catchments larger than 1,000 km^2. Most of the rivers are located in south-eastern China. In addition, the country has about 2,300 lakes larger than 1 km^2 in area, with a total water area and volume of some 71,790 km^2 and 709 billion m^3, respectively. The average annual precipitation amounts to about 6,200 billion m^3, equal to 648 mm, which is 20 per cent less than the world average. Runoff amounts to 44 per cent of total precipitation; that is, China has an annual runoff of 2,712 billion m^3, equal to 284 mm. In a moderately dry year (1 in 4 recurrence) runoff falls to 2,549 billion m^3, while in an extreme drought year (1 in 20 recurrence) runoff amounts to only 2,359 billion m^3. The total average groundwater resources are estimated at 829 billion m^3, but the estimated overlap between runoff and groundwater is 718 billion m^3, giving an average total volume 2,820 billion m^3 of water resources. Total annual water resources amount to 2,840 billion m^3, accounting for one-sixth of the global total; however, per unit area is 83 per cent of the world average. The per capita amount of water resources is 2,210 m^3, equivalent to one third of the world average. Water resources per mu (cultivated land) amount to 1,440 m^3, equivalent to about half of the world average. Water resources are, in fact, scarce in China.

Although the total amount of water is more than adequate, distribution is uneven, in terms of both location and occurrance. Water distribution does not match land resources and fertility allocation. In southern China, total area is 36 per cent of the country, population is 54 per cent, cultivated land is 40 per cent and GDP is 56 per cent, while the water area is 81 per cent of the total. In northern China, the area is 64 per cent, population is 46 per cent, cultivated land is 60 per cent, GDP is 44 per cent, but water amounts to only 19 per cent of the total. In the Yellow, Huaihe and Haihe River basins, the total area is 15 per cent, cultivated land is 35 per cent, population is 35 per cent, GDP is 32 per cent, but water is just 7 per cent of the total, and the per capita water resources is only 457 m3. In this region, there is the most critical conflict between water supply and demand. The monthly distribution of the flow of the rivers closely follows the pattern of rainfall: about 60 to 80 per cent during the wet season, especially in northern China, where rainfall is more concentrated. The two processes of natural water inflow and water demand do not occur in harmony. Water is an important factor affecting balanced development among the different regions, populations, resources and environments in China. Despite the uneven distribution of water resources, they are able to support economic and social development in China through the optimization of water resources allocation, both in terms of location and occurrance.

Scope of the Study

Strategic planning and management (SPM) aims to maximize the benefits to China through its water resources planning and management function. This study considers the component plans of SPM in China as specified in the draft policy, including:

- The principles of water resources management
- Water resources development and management
- Public involvement
- Financial support for water resources development and management
- Water allocation and quality management
- Data and information management
- Capacity-building and human resources improvement

SPM is carried out on the basis of a careful review of past development. In particular, it considers the issues related to the Haihe and Huaihe River basins, including flooding and water logging, water shortage, ecological deterioration and water pollution. The formulation of the development and management strategic plans follows a consultative process in order to obtain the views of the stakeholders as well as the Government of China and its agencies.

Water Resources Potential and Challenges in the Haihe River Basin

Although the Haihe River basin is facing a water resources shortage, it has much potential for supporting socio-economic development by efficient water usage and allocation (e.g., diversion), particularly in the irrigation and water supply subsectors. So far, 32 large-scale reservoirs and 190 medium-sized reservoirs have been constructed within the Haihe River basin, covering more than 85 per cent of this mountainous basin area and providing a total reservoir volume of 29.4 billion m^3. More than 18,000 water diversion and pumping projects have been completed. The large projects include the Beijing-Miyun Water Diversion Project as well as the Water Diversion Projects from (a) the Luanhe River to Tianjin (and Tangshan), (b) the Qinglong River to Qinhuangdao, (c) the Yellow River to Hubei Province, (d) the Yellow River to the northern part of Shandong Province and (e) the Yellow River to the northern part of Henan Province.

A total of 1,200,000 motor-pumped wells have also been constructed.

In 1998, the actual gross water supply within the whole basin was 43.2 billion m^3, including 26.5 billion m^3 of groundwater (rated at 61 per cent), 11.2 billion m3 of local surface water (26 per cent), 5.1 billion m^3 of Yellow River diversion water (12 per cent) and 400 million m^3 from other sources (brackish water, seawater and treated wastewater) (1 per cent). When classified according to the different water consumption sectors, 2.6 billion m^3 was for urban domestic (6 per cent), 140 million m^3 for improving city river and lake environments (0.4 per cent), 7.05 billion m^3 for industry (16.3 per cent) and 33.41 billion m^3 for agriculture (including irrigation, rural living, forestry, livestock and fisheries) (77.3 per cent).

However, water resources exploitation is at a very high level: groundwater is over-extracted, downstream reaches are dry, and water diversion is needed from outside the basin. In addition, several other problems face the water sector, including unusual rainfall patterns in some years, high evaporation rates, flooding and drought in the basin.

Being an over-developed river basin with limited water resources, the most important challenges facing the Haihe River Basin Water Resources Commission (HRWRC) in carrying out its coordinating role include:

(a) The need to strengthen the legal framework for effective and harmonious integration of water resources management, development and protection activities into the socio-economic development process of the basin, especially to meet national priorities;

(b) The need to enhance and consolidate the existing systems as well as provide a foundation for operating, maintaining and rehabilitating facilities safely, reliably and efficiently in order to protect the investment for public benefit; and

(c) Prioritize the capacity-building needs in order to enhance the organizational capacity and effectiveness of the water resources coordination system.

Water Resources Potential and Challenges in the Huaihe River Basin

The Huaihe River basin is situated in eastern China at longitude 112 - 121 east and latitude 31 -36 north, between the Yangtze River and the Yellow River with a total area of 270,000 km^2. The runoff of surface water changes greatly from year to year. The mean quantity of surface water resources is 66.1 billion m^3, while groundwater totals 40.6 billion m^3 and net water resources amount to 96.1 billion m^3. In 2000, withdrawals of water resources totalled 61.2 billion m^3, with exploitation reaching 63.7 per cent.

The water in the Huaihe basin is heavily polluted. The stretches of the river with a water quality worse than class IV is 69 per cent of the total river length, causing water scarcity in the affected areas. The Huaihe basin is also threatened by severe flooding during the monsoon season.

Summary of Goals Linked to the Socio-economic Development Goals

a. Overall Socio-economic Development Context of China and the Haihe River Basin

China is currently undergoing a dual transition from a rural to urban/industrial society and from a command to a market economy in order to integrate into the global economy and promote economic growth. Forces promoting this transition are political and macroeconomic reforms implemented through policy and legal decrees in the areas of (a) state-owned enterprises (SOEs), (b) finance, (c) decentralization, (d) labour markets, (e) government and (f) trades.

The implementation of these reforms is affecting social trends including (i) changes in population growth, (ii) urbanization, (iii) rising income and income disparities, and (iv) rising consumption of goods and services. Those social trends and economic reforms combined with natural and historical conditions have various impacts on water resources and the institutions that manage them. Complex technical, social and management solutions are required for addressing the growing water shortages and pollution.

According to the tenth Five-Year Plan (2001-2005), the economic growth rate is expected to reach 7 per cent; that is, GDP will be RMB 12,500 billion in 2005, in which the shares of agriculture, industry and service sectors will be 13 per cent, 51 per cent and 36 per cent, respectively. The population growth rate is expected to slow to less than 0.9 per cent in 2005 with a total population of some 1.33 billion.

The population of the Haihe River basin is 118 million, representing 10 per cent of the national total, of which the urban and rural populations account for 24 per cent and 76 per cent, respectively. The average density is 371 persons/km2, which is 3.47 times the average national density. Moreover, there are large differences in density between provinces (cities) in the basin, ranging from 56 persons/km2 to 784 persons/km2. At present, there are two municipalities directly under the central Government, five provinces and an autonomous region, including 65 regional administrative districts (Beijing and Tianjin comprise 36 districts and counties), all of which are at the regional level and 235 administrative districts at the county level.

Cultivated land in the Haihe River basin totals 163 million mu, which is 11 per cent of the national total. Of this area, dry land accounts for 158 million mu and irrigated fields for 4.79 million mu. The effective irrigation area is 102 million mu. Cultivated land per capita is 1.38 mu, equal to the national average value. From 1949 to 1993, cultivated land decreased by 23.2422 million mu, hence 14 per cent of present cultivated land, or an average annual decrease of 520,000 mu, hence 0.3 per cent of present cultivated lands. Therefore, cultivated land should be valued. Moreover, the remaining undeveloped 40 million mu can be developed.

In Haihe, the total value of agricultural output is RMB 97.5 billion, while total food yield is 45.4 million metric tons, accounting for 10 per cent of the national total. Food per capita is 385 kg. The gross value of industrial output is RMB 644 billion (of which RMB 270 billion is from village/town enterprises that account for 36.4 per cent), representing 15 per cent of the national total. The total value of industrial and agricultural output combined is RMB 742 billion, while overall per capita income is RMB 6,285.

The Haihe River basin is rich in mineral resources such as coal, petroleum, gas, steel, aluminium, gypsum, graphite and sea salt. For example, coal reserves are some 202.6 billion mt, which is nearly 30 per cent of the national coal reserves; annual extraction is 28 million mt, or 20 per cent of the national figure. Moreover, the Huabei, Dagang and Shengli oilfields as well as part of the Zhongyuan oilfield are in the Haihe River basin, with 1.5 billion mt of oil reserves and 36 million mt of annual output.

The Jing-Guang, Jing-Jin, Jing-Ha, Jing-Hu, Jing-Bao and Jing-Yuan railway lines cut through the Haihe River basin. According to available but incomplete statistics, in Beijing City, the volume of passenger traffic is 110 million persons per year, while the volume of freight traffic is 460 million mt. Along the Bohai Sea, there are the ports of Qinhuangdao, Tanggu and Huanghua, among others, with 109 berths and a handling capacity of 140 million mt. The air transport centre is Beijing, with airports in Tianjin, Shijiazhuang, Qinhuangdao, and Xingtai and Changzhi, accounting for 4.16 million passengers and 2.98 million mt of freight traffic annually.

The vast Huabei Plain, with its good climate and abundant sunshine, is one of the main grain producing regions. The food crops comprise mainly wheat and corn, plus Chinese sorghum, rice, millet and beans. Industrial crops are cotton and oil plants. In 1993, the total value of agricultural output was RMB 94.47 billion, and the average grain yield per mu was 292 kg. Currently, productive power is weak with the exception of a high output in the alluvial plain and Yellow River diversion irrigation region. The Haihe River region is also rich in fruit, and its forestry productivity and stock-raising can be developed further, combined with water and soil conservation. Moreover, there are 3 million mu of saline-alkaline wasteland and many reservoirs, and the low-lying lands should be developed.

b. Opportunities and Threats for Water Resources Development from the Perspectives of the Haihe River Basin and Related Provinces

With more than 18,000 water diversion and pumping projects having been completed, including large projects such as the Beijing-Miyun Water Diversion Project, the Water Diversion Project from Luanhe River to Tianjin municipality and Tangshan City, and the Water Diversion Project from the Yellow River to the northern part of Henan Province.

In 2000, the actual gross water supply for the basin was 40 billion m3, including 26.3 billion m^3 of groundwater (about 65.8 per cent), 13.6 billion m^3 of surface water (including diversions, accounting for 34 per cent) and 100 million m^3 of water supplied from other sources (brackish water, seawater and treated sewage). Overall water consumption by sector is 5.2 billion m^3 for domestic consumption (some 13 per cent), 6.6 billion m^3 for industrial uses (some 16.5 per cent) and 28.1 billion m^3 for agricultural uses, accounting for 70.3 per cent of the total.

With this demand/supply balance, the annual water flow rate to the sea decreased from 24 billion m^3 in the 1950s to 6.7 billion m^3 in the 1990s. The reduced flow has caused serious sedimentation in the river mouths and estuaries.

For the past 10 years, basic water demand for socio-economic development could only be satisfied by over-extracting by some 5-7 billion m^3 of groundwater and re-using nearly 3 billion m^3 of untreated sewage water. The shallow groundwater over-extraction area is presently about 44,000 km^2; however, in some 10,000 km^2 of the sediment area the shallow groundwater resources have dried up.

The deep groundwater over-extraction area is 56,000 km^2. The total over-extraction area of both shallow and deep groundwater resources in the whole basin is estimated to be 89,000 km^2, accounting for 69 per cent of the plain area of the basin. Consequently, environmental and geological problems, such as land subsidence, have been induced in the areas of extreme over-extraction.

Due to the excessive utilization of surface water resources, many rivers dry up in most seasons or are now dry throughout the year. In addition, over-discharging of pollutants into the rivers has left much of the remaining water seriously polluted. An example from 1998 best illustrates the serious water situation. During that year, total precipitation was 175.3 billion m^3 (equivalent to 551 mm in depth) and the gross volume of water resources amounted to 35.5 billion m^3, which represents an average year.

However, the water resources utilization ratio was 90 per cent, and the total water flow to the sea (including untreated sewage) was 4.3 billion m^3.

c. Vision Statement of the Haihe River Basin Water Resources Commission

By consulting representatives from HRWRC, the Ministry of Water Resources and some of the provinces concerned, the following vision statement for HRWRC has been determined: "Towards a most modern, reliable and efficient water resources management system to protect, manage and develop the World's largest complex water transfer system to achieve sustainable well-off life in the Haihe River basin."

The highest organ of the Government is the National People's Congress (NPC). Of nine special committees established by NPC, two are of particular importance to the water sector: the Environment and Resources Protection Committee, and the Agricultural and Rural Affairs Committee. The Chinese People's Political Consultative Conference (CPPCC) is the highest advisory body in the country, meeting on an annual basis and consisting of more than 2,200 representatives from all social sectors. Its primary function is political consultation, democratic supervision and participation in, and deliberating on State affairs.

Executive power is exercised by the State Council. Under the State Council are the functional organizations such as ministries and commissions. According to the official government structure, provincial level departments and bureaus of water conservation, and community level water conservation stations and water users associations are vertically integrated with the Ministry of Water Resources. However, they also report and depend on the provincial governments especially in relation to administration of laws and importantly for funding. This form of decentralization may be defined as "delegation" where the central government transfers responsibilities for decision-making and administration of public functions to local governments.

The "principled laws" that guide the activities of the ministries in water resources management or environmental protection are based on well-established scientific doctrines and China's own blend of water administration law, enforcement and jurisdiction. These well-meaning principles (such as water allocation based on watershed or polluter pays principles) increasingly conflict with the economic interests of the provinces. Those administrative bodies (whose financial viability depends on provincial budgets) empowered by law to enforce rules and regulations at the local level (such as water bureaus) are often under pressure to act in the interests of the local government to the detriment of sound water or environmental resource management and planning. Thus, vertical relationships between ministries, departments, bureaus etc. have been challenged by stronger alliances between the local governments and these local administrative bodies.

In an effort to improve holistic planning for water resources, coordination among ministries, commissions and lower level government entities is effected through "leading groups" (coordination committees) or interdepartmental agreements. The main functions of the leading groups are to:

(a) Examine and approve the comprehensive planning of large river basins

(b) Examine and approve the important principles and policies of national water and soil conservation work as well as the important problems of key control projects

(c) Resolve major water resource allocation problems among various sectors

(d) Handle and coordinate major water affair conflicts between provinces

The only standing inter-ministerial committee in the water sector at the national level is the State Flood and Drought Control Relief Headquarters. During times of emergency, this office reports directly to the responsible vice-premier. A temporary coordinating committee may be established for a particular purpose (e.g., related to a specific basin or programme or issue), and if so, is chaired by the appropriate ministry. Provinces may also establish leading groups for specific programmes or issues within their jurisdiction, although again this practice may now be discouraged.

As at the central level, provincial planning commissions are responsible for reviewing and approving programmes and projects, and may play an important role in resolving wider coordination difficulties. In practice, there is considerable overlap between the different ministries, and inconsistent and fragmented responsibilities are a major issue. While legislation calls for integrated and comprehensive management, in practice, different ministries and agencies are primarily responsible for implementing specific acts of legislation; thus, inconsistencies and conflicts arise between different agencies in the exercising of their respective mandates. This is probably why resource conflict mediators and interprovincial water allocation planners such as leading groups were set up in the first place. However, the political process of decentralization (which has undoubtedly helped to boost China's economic growth) has taken away this high-level planning perhaps because it is viewed as being too centralized.

In addition, with the intensification of competition between provinces due to (a) increasing economic activity, (b) population growth and (c) new imperatives of regional responsibility systems, local governments have had to acquire planning/management rights to natural resources in order to sustain economic growth. Micro planning and management at the regional level, limited by administrative boundaries, is thus the current reality in the water sector. Certainly, ministries have retained significant input into planning and management, but as central government revenue and public expenditures have shrunk, but they are constantly constrained by the limited resources available for public investment in national priorities and nationwide externalities such as large water transfers, wastewater treatment plants etc. These key projects may have a far greater significant impact on growth across provinces than their counterparts in each province. In addition, given the nature of water resources as a unitary good, multiple planning and regional management may have negative effects on economic growth.

Apart from inter-ministerial leading groups, other mechanisms designed to improve water resources management include the river basin management groups.

The Water Law recognizes the basin as the logical context for devising solutions to water resources management problems. How best to coordinate activities based on hydrological boundaries with those based on administrative boundaries is a perennial issue in water resources management. Among the issues that have arisen are the following:

(a) The River Basin Commissions (RBCMs) are commissions in name only, having no separate governing board or corporate status. RBCMs are departments of the Ministry of Water Resources and perform those functions that the ministry delegates to them. RBCMs find it difficult to enforce provisions of basin plans on other sector ministries and provincial governments, and the functions that they perform overlap with activities undertaken at the provincial and local levels;

(b) RBCMs in principle help resolve conflicts between jurisdictions and sectors, and ensure that multiple uses are served according to established priorities. However, these functions are hampered by an absence of formal agreements on interprovincial water allocation, pollution limits, and other matters. Only in the case of the Yellow River has water been allocated among the provinces and, even here, provisions for varying river conditions are inadequate. RBCMs may propose and (to the extent that they undertake the real-time management of multipurpose facilities) enforce allocations; however, most storage and diversion facilities are controlled by local or sector entities, and these may be operated in ways that are inconsistent with the provisions of RBCM plans. Only during times of flooding do RBCMs exercise predominant authority over other entities under the State Flood Control and Drought Relief Headquarters;

(c) The administration of land, groundwater and water quality is typically handled by agencies at the provincial and local levels. The transfer of aspects of groundwater administration to the Ministry of Water Resources and the provincial water resource bureaus may facilitate integration of groundwater and surface water administration, but conjunctive management as well as the integration of land and water, and water quantity and quality, all remain illusive within the basin context. Little effort is made to consolidate plans covering each aspect within comprehensive county, provincial or basin plans; and

(d) In principle, RBCMs prepare basin development and operating plans, and undertake other tasks in full consultation with the provinces, sectoral ministries and other stakeholders. In practice, there are few formal consultation mechanisms, and the main directives and decisions affecting RBCM activities are received vertically from the Ministry of Water Resources.

Legal and Institutional Framework

a. Legal and Institutional Context

In 1954, the first Constitution was promulgated and was the foundation document for the socio-economic management of the country. It provided the basis for a multi-sector economy as well as the conditions for the education, culture and health of the Chinese people. Thereafter, the second Constitution was enacted in 1975, the third in 1978 and the fourth in 1982. Since 1982, the Constitution has been revised four times in accordance with the changing domestic, regional and global environment. In March 2004, the Constitution was revised in order to meet the real needs of the Chinese people in the current economic environment.

The first Water Law was formulated by the State Council and adopted by the National People's Congress Standing Committee in 1988. It was revised in 2002 as both the water resources conditions and water utilization situation had changed considerably. The Water Law sets out a legal framework for development of the water sector. The development of further legislation or a decree on subsector activities is urgently needed as well as the necessary legal documents to make the law effective. External assistance is also required by water subsector agencies in the drafting of regulations covering the operation and enforcement. A prerequisite for success in this area is capacity-building throughout the water sector, which traditionally focuses on development rather than management and regulation.

In addition, other laws related to water include the Flood Control Law, Water Pollution Protection Act, Water and Soil Conservation Act, Environment Protection Act, Land Management Act. The Flood Control Law was enacted to (a) prevent and control flooding, (b) alleviate calamities caused by flooding and water logging, (c) maintain the safety of people's lives and property, and (d) safeguard the smooth progress of construction modernization.

The Water and Soil Conservation Act was formulated for the purposes of (a) preventing and controlling soil erosion, (b) the protection and rational utilization of water and soil resources, (c) the mitigation of flood, drought and sandstorm disasters, (d) the improvement of the ecological environment and (e) the development of production. The Water Pollution Protection Act was enacted in 1996 to (a) control water pollution and protect the environment, (b) guarantee the health of the people and (c) promote the modernization of construction.

The Water Law grants the State Government the right to apply water licences and fees for water use. The Water Administrative Department of the State Council is in charge of the organization and implementation of water licensing and fees.

The State Government has adopted an approach that combines river basin management with district management. The Water Administrative Department of the State Council, which is supported by national laws and administrative decrees, is responsible for the integration of management and supervision of national water resources including important rivers and lakes defined by the State. It is also in charge of the exploitation, utility, efficient use and protection of water resources. The Ministry of Water Resources is responsible for integrated management and supervision of national water resources.

The provincial, municipal and county level governments are in charge of the integrated management and supervision of water resources within their districts, according to the Water Administrative Department.

The relevant departments of the State Council are responsible for the exploitation, utility, efficient use and protection of water resources. The provincial, municipal and county level governments are in charge of the exploitation, utility, efficient use and protection of water resources within their districts.

HRWRC, as an administrative agency of the Ministry of Water Resources, is authorized to undertake water resources management and supervision and is supported by national laws, administrative statues and the Ministry of Water Resources.

b. Current Situation and Perspectives of the Legal and Institutional Framework

The Ministry of Water Resources is a Water Administrative Department of the State Council. Other relevant departments include the State Environment Protection Administration, the Ministry of Land and Resources, the Ministry of Construction, the State Forestry Authority, the National Development and Reformation Committee, the Ministry of Communications, the Ministry of Public Health, the Ministry of Finance and the State Meteorology Authority. The Ministry of Water Resources grants its seven river basin administrative institutions the authority to manage the water resources within their regions.

As a subordinate body of the Ministry of Water Resources in the Haihe River basin and northern part of Shandong Province, HRWRC is empowered to administer the above-mentioned areas with regard to (a) managing the water resources and river course of the local basin, (b) control and develop important water conservancy projects, (c) undertake planning, management, coordination, supervision and advancement of river control, and (d) ensure all-around development, utilization and water resources protection.

HRWRC comprises a water administration and management department, an hydraulics and hydropower administrative department (flood prevention and drought prevention office), a department of water and soil conservation, a water resources protection bureau, an hydrology office, the Haihe River lower reaches administration bureau, the Zhang-Wei south channel administration bureau and the Zhanghe upper reaches administration bureau.

It should be noted that there is no clear river basin law or regulation that sets out standards for how a river basin administrative agency should carry out water resources management and supervision, despite the functions explicitly defined by the Water Law.

Development of clear legislation, regulations and guidelines is one of the primary means by which HRWRC can play its coordinating role within the water resources sector of the Haihe River basin. The process must be handled in an open and cooperative way, involving the stakeholders at the central, provincial and local levels. HRWRC has already gained some experience in this aspect, but further capacity-building for policy analysis and development is required. Legislation and institutions need to be supported by detailed implementation plans that indicate agency responsibilities and, where necessary, further capacity-building to allow implementation to be successfully carried out.

c. Strengths and Weaknesses of the Current Legal and Institutional Framework

Enforcement measures and detailed rules and regulations are lacking for the Water Law, Law of Water and Soil Preservation, Law of Environment Protection and Flood Control Law. In addition, some connection problems exist between local rules and regulations and the administrative management system. Some local rules and regulations are not in accordance with those of the State, and there are no special procedures for the inspection and coordination of enforcement measures at the local level. When enforcement measures are introduced, prior opinions are not sought from HRWRC, which makes it difficult for HRWRC to maintain control.

Although the water legislative position, legislative supervising authority and legal duties of HRWRC are defined in the existing Water Law, it is not specific. There are no effective management measures for ensuring the performance of the river basin plan, such as water resources development, allocation, supervising management, basin control, project approval, and engineering investment. Even if a project is in violation of the law, RBCMs are not granted the authority to take effective measures for preventing illegal activities, and the performance of the basin plan also fails to be effective. Thus, it is clear that the corresponding detailed rules and regulations for water resources management need to be formulated, such as those concerning water quantity and quality, and flood control. At present, the following problems require urgent attention:

- The fundamental nature, position and responsibilities of RBCMs, and authority to enforce the law.
- The relationship between the river basin management and administrative sectors.
- The management principles of the river basin and the basic management system.
- Policy measures for promoting the comprehensive development of the river basin.

Despite its title, the existing River Basin Council is only an agency of the competent water administrative department of the State Council. In fact, management is not carried out by the officials of the relative departments and sections. At present, following the establishment of the flood control headquarters and water resources protection agency, a great deal of successful experience has been gained. Although the RBCM departments are mainly administrative institutions and deal with few enterprises, their institutional expenses are allocated by the financial department; however, the amount provided is less than the half of actual expenditures. Without sufficient financial support, an RBCM will be unable to handle its responsibility. While an RBCM is in charge of some water supply and hydropower projects it has no authority to decide on the water and electricity prices. In effect, the charges for water and electricity are much lower than their costs; however, neither compensation for resultant losses nor starting and development funding are provided by the Government.

The management model that takes the river basin as a basic management unit for water resources will be unable to achieve efficient impact unless the management model used for water project funds provided by the State is the same as that used for a river basin. Together with the reformation of the State planning system, financial support by the central Government for such projects will be greatly reduced. The current institutional problem in the water sector is related mainly to the lack of coordination between agencies within the sector and with those of other sectors, as well as the loose communication and coordination between the national agencies and their provincial counterparts. The first step is to assess the present institutional set-up at the provincial and river basin levels, and to conduct a diagnostic study of each targeted river basin. The findings of these studies could be used as a basis for designing a river basin management body tailored to the development mix in each basin.

d. Haihe River Basin Water Resources Commission Mission Statement

The mission statement suggested by the water sector is "Protect, manage and develop water resources to achieve good living standards in the Haihe River basin". As with the overall vision statement, the words have specific meanings. Protect refers to the need to ensure the continued supply of high-quality water resources. Manage refers to the water resources management activities, including access to water resources and the right to utilize water resources, and managing those institutions with responsibilities related to water resources. Develop refers to converting water resources to new uses and to a mature condition for the purpose of improved quality of life.

The Circular of Functions and Duties of the Haihe River Water Resources Commission, No. [1994]24, published on 16 March 1994 by the Ministry of Water Resources, defines the rights and duties of HRWRC. Although broad, the list of rights and duties is mainly directed towards actions, such as "planning, managing, coordinating, supervising and serving" as follows:

- Supervise the enforcement of the Water Law and the Water and Soil Conservation Law, and formulated policies and regulations for the river basin;
- Work out medium- and long-term water resources development plans for the Haihe River basin, formulate a comprehensive plan and specialized plan together with the relevant departments and provincial government, and supervise enforcement of the plan after approval;
- Organize water resources monitoring and assessment in an integrated manner, formulate a long-term water supply and demand plan and a water distribution scheme, and undertake the supervision and management of water resources protection;
- Manage the river, lakes, river mouth and key river course in the basin, and examine construction projects;
- Formulate flood control plans for the river basin, supervise the flood control plans of trans-provincial rivers, coordinate the daily work of flood and drought control in the river basin, and supervise safety and construction in flood areas;
- Mediate water affairs among departments and provinces;
- Organize prevention, supervision and all-around control in areas where loss of water and erosion of the soil is serious;
- Examine the project suggestion report, feasibility plan and preliminary design of projects directly under the central authorities as well as local joint project;
- Be responsible for the comprehensive development of the river basin and the management of important trans-provincial water conservation projects;
- Direct the work of rural and urban water conservation, water conservation project and electrification project management; and
- Undertake other work as required and authorized the Ministry of Water Resources.

Strategic Plan Goals

a. Overall Haihe River Basin Water Resources Commission Goal

The overall HRWRC goal is to achieve consensus building and sustainable development of the water resources in the country, particularly in the priority areas identified by the Government. Within the context of this overall goal, an attempt will be made to achieve the following three integrated water and resources management (IWRM) main mission goals:

- IWRM Goal 1: To improve and modernize the legal and technical framework for managing, developing and protecting water and related resources in order to meet the needs of current and future generations;
- IWRM Goal 2: To enhance and consolidate the existing systems and foundation for operating, maintaining and rehabilitating facilities safely, reliably and efficiently in order to protect public investment; and
- IWRM Goal 3: To enhance the organizational capacity and effectiveness of the water resources coordination system.

The three mission goals are linked to the three priorities of the country mentioned at the beginning of the report, in terms of ensuring high economic growth, poverty alleviation and a strong framework for sustainable development.

b. Haihe River Basin Water Resources Commission Mission Goal 1

To achieve this mission goal, the Ministry of Water Resources will work in cooperation with all key ministries and stakeholders towards effectively managing, developing and protecting water and related resources for agricultural, municipal, industrial, rural, hydropower, recreational, fisheries and wildlife purposes.

The Ministry of Water Resources and the Haihe Basin Commission will pursue coordinating roles aimed at improving water and related resource management, development and protection activities. The framework for such coordinating roles would include:

(a) Legislation to improve integrated and comprehensive water resources management, including flood control, irrigation, urban and rural water supply, water pollution control and treatment;

(b) Legislation to strengthen the authority of the Haihe River Basin Commission to implement integrated and comprehensive water resources management. The Haihe River Basin Commission should be given more implementation authority to ensure sustainable water use, rather than just design and planning;

(c) Formulating comprehensive water resources plans, including surface water and groundwater, water quantity and quality, water use and protection, economic and environmental water use etc., which should get approval from the State Council and should act as guidelines of management activities;

(d) Formulating policy and strategies to improve the water resources planning systems of the country, with emphasis on river basin water resources planning systems;
(e) Formulating policy and strategies to improving the efficiency of water use; and
(f) Promoting integration of applied sciences and technology into water resources management.

c. Haihe River Basin Water Resources Commission Mission Goal 2

Under this mission goal, the Ministry of Water Resources and the Haihe Basin Commission will focus on operating, maintaining, and rehabilitating existing water resources facilities so that they continue to provide project benefits. These facilities provide power and water supply delivery systems that serve agricultural and municipal users, in addition to recreation, fisheries and wildlife benefits, and flood control.

Ensuring that the facilities are safe, cost-effective and reliable will assist the agencies concerned to operate them as effectively and efficiently as possible in order to provide project benefits while also protecting public health, sustaining environmental values, and providing timely and economical services to customers. This will protect the public investment.

To enable this goal to be achieved, the necessary legislation for ensuring safe and reliable water related services should be promulgated. At the same time, certain agencies should be endowed with the responsibility for maintaining the safety and reliability of such facilities.

d. Haihe River Basin Water Resources Commission Mission Goal 3

In order to achieve the above mission goals, HRWRC needs to be strengthened to make it more effective. The strategies for achieving the goal should include:

(a) The formulation of a reform plan for the Haihe Basin Commission to adapt to the increase in authority and the changed situation, with the focus on managing water resources in an effective and integrated manner. The reform plan will be presented to the Government for approval;
(b) The establishment of an information management system to facilitate management;
(c) The preparation of five-year Action Plans, based on the priority water resources management issues at the national level and river basin levels; and
(d) The preparation and implementation of a capacity-building plan for HRWRC, based on the requirements for achieving the Action Plan.

Implementation, Monitoring and Evaluation

a. Identification of Indicators and Benchmarks for Performance Measurements, and the Expected Timeframe and Necessary Resources

Implementation would be carried out based on the clarification of the strategic planning and management process.

(i) Human Resources Development

It is necessary to strengthen the River Basin Commission, set up a water resources coordinating committee, and develop the right skills and competencies to match the strategy, giving recognition to the fact that the Chinese people are the best source of ideas for their future. Moreover, confirmed leadership, creative teamwork and organizational culture are important to the successful implementation of strategic plan management.

Consequently, it will be necessary to:

(a) Build an organization around the critical strategic elements;

(b) Mobilize and allocate resources to match the strategic plan;

(c) Establish administrative systems, policies and procedures in accordance with the strategic plan;

(d) Develop a personnel reward system linked to achievement and the overall strategy; and

(e) Recognize that the most important job of leadership is to establish and instil confidence in the strategic plan management, and to ensure that everyone in HRWRC follows it.

Because of the complexity of strategic planning, it will be necessary to seek financial support, both at home and abroad. Only if HRWRC treats fund-raising as a fundamental part of the operation, and approaches it in a professional and systematic way, will strategic planning have any chance of becoming financially secure. As the Ministry of Water Resources has appointed HRWRC as the leading institution in the promotion and coordination of water resources development for the benefit of the river basin population, its duty is to treat fund raising not as an auxiliary issue, but as an essential component of the overall operation.

Financial security means that funding must be sought and obtained in a flexible manner from diverse sources at various levels, including the central Government and, in particular, the Ministry of Water Resources, local governments, communities and associations, involved enterprises as well as international organizations.

It is clear that a subsidy from the central Government will play a crucial role, while local governments would provide of counterpart funds. Financial assistance contributed by other institutions or organizations will be in a variety of forms, such as cooperative scientific research, transfers of project results at a preferential price or even free of charge. As for seeking overseas funding, this would take a similar course. In addition, HRWRC responsibility would include the identification of reasonable targets for funding and in-kind services during the process of strategic implementation as well as deciding on how the funds and services should be used.

With regard to financial management, HRWRC is accepted as the core authority in charge of all relevant issues. Its functions are clearly defined by the central Government, particularly the Ministry of Water Resources, and bolstered by local government at different levels.

The major duties performed by HRWRC are to (a) put forward the overall budget needed, (b) determine the framework of fund-raising and clarify the contribution percentages from different financial resources in the foreseeable future, and (c) establish and carry out efficient financial monitoring systems over the strategic planning period.

(ii) Action Plan

Three levels of action by HRWRC will be necessary, as described below.

(1) *First Level:* The first level includes the reasonable allocation of water resources, water conflict reduction and the establishment of a consolidated information system.

(a) Reasonable allocation of water resources. Water resources include surface water, groundwater, treated wastewater, brackish water and fresh seawater etc. Water use involves many aspects, such as domestic urban and rural usage as well as production and eco-environment usage. Optimized water allocation will be achieved via water allocation, coordination of water supplies and usage, unification of water quantity and quality, and real-time regulation and preference of the benefits. The mutual relationship of economic structures, economic development and water resources must be handled correctly via administrative, economic, technical and legal measures. Reasonable water resources allocation can ensure that limited water resources produce full benefits.

(b) Water conflict reduction. The River Basin Commission should coordinate the relationship between the provinces concerned, and between the different water-use sectors. By establishing an information sharing system, such a relationship can be formed based on mutual respect and trust. Using market mechanisms and establishing a compensatory mechanism are also important ways of reducing water conflicts.

(c) Establishment of a consolidated information system. On the one hand, the province will provide basic hydrological and water use information to the River Basin Commission. On the other hand, the River Basin Commission should provide information to the province and the public. This system will not only reduce water conflicts but also improve water use efficiency.

(2) *Second Level:* As detailed below, the second level of action includes water conservation and improved water-use efficiency, the use of seawater in coastal areas, the construction of a south-to-north water diversion project and the protection of groundwater resources.

(a) Water conservation and improved water use efficiency. At present, the industrial water saving level in Haihe River basin is very high, and the average water consumption quantum for every RMB 10,000 output value has been reduced from 160 m^3 in the early 1990s to 51 m^3. The water reuse rate is about 80 per cent, which is 40 per cent higher than the average national level. The efficiency of industrial water consumption can only be improved by adjusting the industrial structure and arrangements, limiting the high level of water consumption in production, adopting water-saving technology and gradually increasing the ratio of new technology industries.

Use of irrigation water in the basin is about 30 billion m^3, which is 70 per cent of the total water supply. The comprehensive irrigation quantum is some 296 m^3/mu. The water-saving irrigation area is 40 per cent of the total irrigation area, and it is planned to increase the percentage to 70 per cent in 2010. Together with the process of urbanization, domestic urban water use will be steadily increased. Domestic urban water saving will be realized mainly by popularizing water-saving appliances, raising the water price, reducing seepage losses in the running water conveyance pipe network, developing repeated usage and enhancing public awareness of water saving.

(b) Use of seawater in coastal areas. Haihe River basin has an approximate 900-km coastline. Therefore, the use of seawater is an important way of mitigating fresh water shortages in coastal cities in the future. At present, seawater is only used for cooling by the electric and chemical industries in Tianjin, Qinhuangdao and the coastal areas of Shandong Province. The use of freshened seawater lags far behind. Together with the continuous progress in seawater usage technology and the reduction of freshening costs, the use of seawater will have a bright future. Freshened seawater will eventually undergo even wider production.

(c) Construction of a South-to-North Water Diversion Project. The project being established is of strategic fundamental infrastructure for solving the water shortage problem in Haihe River basin. The two largest cities, Beijing and Tianjin, and the northern China plain are intake areas of the East Route and the Middle Route of the South-to-North Water Diversion Project. It is related to the domestic water supply as well as economic development and improvement of ecological environments in those areas. An analysis of the present and future water resources demand and supply in the basin shows that merely developing efficient water use and the potentialities cannot thoroughly change the water shortage situation and the worsening trends in the ecological environments. The water diversion project is necessary to meet water demand.

(d) Protection of groundwater resources. The groundwater resources of China, especially in the Haihe River basin, play a vital role in water resources that have served China well. The Haihe River basin has large areas where both shallow and deep aquifers are greatly overexploited. Approximately 28 per cent of the total groundwater used is from deep confined aquifers where the recharge is very limited. The heavy overexploitation of groundwater resources causes huge environmental damage - subsidence, seawater intrusion, salinity degradation, pollution, and falling groundwater levels. The economic consequences of these effects are great. The serious pollution of surface water and surface water scarcity has driven the increasing use of groundwater during the past few decades. Groundwater quality, especially in the deep aquifers, is far better than the surface water quality. However, increasingly serious pollution of the shallow aquifers is now encouraging the growing use of deep groundwater resources. The very serious and largely irreversible falling groundwater levels throughout the northern China Plain demands a major programme of groundwater management planning to reduce groundwater use to sustainable levels.

The groundwater management strategy proposed will require a huge effort; however, the consequences of not doing so will have serious long-term side-effects, such as effectively destroying the groundwater-dependent agricultural base, massive subsidence and seawater intrusion, virtual elimination of groundwater as a water source for many cities and countless households, and the loss of "insurance" water for future generations. The fundamental objective proposed is to reduce groundwater use to sustainable levels by 2015. A broad range of technical, institutional and management actions are required to achieve this goal. The key actions required are:

(a) The definition of all significant groundwater usage areas as Groundwater Management Units and the determination of the sustainable yield;
(b) Preparation and implementation of groundwater management plans as per the programme described here;
(c) Allocation licensing to be linked to sustainable yield assessment;
(d) Allocation licensing only to be undertaken by one department;
(e) Licensing of well construction drillers;
(f) A River Basin Groundwater Database needs to be developed;
(g) A groundwater pollution prevention strategy needs to be prepared;
(h) A review by the Ministry of Water Resources of the adequacy of regulations required to implement the proposed groundwater management reforms;
(i) The introduction of realistic groundwater prices; and
(j) The preparation and introduction of a major education programme on groundwater processes and the need for groundwater management.

(3) *Third Level:* The third level comprises the improvement of water resources institutional management, and integrated planning.

The improvement of water resources institutional management should involve:

(a) The establishment of a high-level River Basin Coordinating Committee or Council in the Haihe River basin, chaired by a Vice-Premier with the Ministry of Water Resources as Vice-Chairman, and comprising: (i) the mayors and governors of the constituent municipalities, provinces and autonomous regions, (ii) the Vice-Minister for Water Resources and the Vice-Ministers of SEPA and the other key "water" agencies - the Ministries of Agriculture, Construction, Finance, Communications, the National Development and Reformation Committee, and the State Meteorology Authority; (iii) the Commissioner or Director of the respective RBCM; and (iv) representatives of key civil societies such as water supply companies and water user associations;
(b) The Haihe River Water Resources Commission should be charged with the responsibility for: (i) determining water resources allocations (surface and groundwater) for the provinces (real-time, and dependent on the water held in reservoirs and seasonal conditions); (ii) the development of broad policies and programmes promoting sustainable water resources management, particularly with the respect to flood control and drought relief, groundwater management, water

resources protection and pollution control, and the promotion of increased water-use (especially irrigation) efficiency; and (iii) comprehensive basin development planning;

(c) Granting the River Basin Councils the necessary legislative support and authority to ensure essential coordination and enforcement of its allocation, policy and programme decisions in the municipalities, provinces and autonomous regions;

(d) Confirmation by law of the River Basin Councils as the primary water resources management agencies in the Haihe River basin (with specific flood control responsibilities) to ensure the enforcement of council decisions on the allocations, policies and programmes in consultation with the municipal, provincial and autonomous region governments. It can be strengthened with the appropriate expertise where necessary to provide administrative and technical secretariat support for the River Basin Council; and

(e) Adoption by the River Basin Councils of the strategic planning and management methodology required to determine the policy, programme and project priorities required to accelerate the basins towards achievement of sustainable development. With regard to integrated planning, the River Basin Commission should prepare an integrated planning guide to the water resources development and protection at the province level.

(iii) Mechanisms for reporting, monitoring and evaluation

(1) *Coordination Framework:* The ministries and agencies responsible for the management, exploitation, development and use of water and water resources include:

(a) The Ministry of Water Resources. Responsible for integrated management and supervision of national water resources as well as the exploitation, development, use and protection of water resources in China. It is also responsible for monitoring, control, promotion and reporting on the implementation of activities related to water and water resources.

(b) Ministry of Agriculture. Responsible for the exploitation, development and use of water resources in agriculture.

(c) State Forestry Authority. Responsible for the exploitation, development and use of water resources in forestry.

(d) Ministry of Communications. Responsible for the management, exploitation, development and use of water resources in the field of communications. In addition, the ministry is responsible for collection of hydrological data and hydrographical surveys for navigation.

(e) Ministry of Construction. Responsible for the management, exploitation, development and use of water resources in the municipal areas, town water supplies, urban drainage and wastewater treatment.

(f) Ministry of Public Health. Responsible for the management, exploitation, development and use of water for rural domestic consumption and health care.

(g) State Environment Protection Administration (SEPA). Responsible for ecologic environment protection and wastewater discharge management.

(h) Ministry of Land and Resources. Responsible for hydrogeology surveys.

(i) HRWRC. Empowered to exercise authority of water administration in the above-mentioned areas. Responsible for managing the water resources and river course of the local basin, according to the rule of management in unity and rank. Responsible for controlling and developing important water conservancy projects comprehensively. Ensure efficient planning, management, coordination, supervision and servicing. Undertake river control, and all-around development, utilization and water resources protection.

The above ministries and agencies would need to coordinate with the local authorities in the detailed determination of responsibilities and scope of activities within their sectors.

(2) Licensing of water use and water resources development activities

The State Council has issued Decree No. 119 on licensing water catchments. The Decree determines that the Ministry of Water Resources will implement this licensing system and the different scales of water catchment licences to be issued by water administrative institutions.

The water administrative departments should approve important medium-scale water uses. The Ministry of Water Resources or the River Basin Commissions should be responsible for approving the water catchments of individuals and organizations wishing to use water from the Yangtze, Yellow, Huaihe, Haihe, Luan, Pearl, Songhua, Liaohe, Jinsha and Han Rivers, international rivers, State border rivers, inter-province rivers indicated reaches, provincial border rivers and lakes, inter-provincial regions and water quantities beyond the set quota. The water administrative departments of local authorities should be responsible for approving other water catchments.

Permission for large-scale water resources development activities should be sought from the Ministry of Water Resources, SEPA and NDRC, and final approval given by the State Council. Applications for large-scale water resources development activities should be accompanied by a feasibility study and environmental impact assessment. Smaller-scale water resources development activities should apply to the water administrative departments and local government authorities.

The Ministry of Water Resources and SEPA need to prepare guidelines on the preparation of feasibility studies and environmental impact studies, respectively, in accordance with the Environmental Impact Evaluation Law. The ministry also needs to monitor and control the development of water resources.

All water resource development projects, particularly the construction of dams, irrigation schemes, and the diversion of rivers and streams, should go through a public consultation process with the local population. When a water resource development project is expected to have an impact on the natural or social environment, SEPA, the Ministry of Water Resources and local authorities need to:

(a) Issue a project notification to inform the public through newspapers, public notices and other means and, depending on the type of project, determine the duration for such a notification; and

(b) Organize public hearings in the relevant location. The developer or representative should attend the public hearings and provide a detailed explanation and information on the proposed mitigation measures. All costs associated with the public hearing will be the responsibility of the developer.

(3) *Monitoring and Evaluation:* The Ministry of Water Resources, HRWRC and local authorities are responsible for the monitoring and evaluation of the implementation of measures provided in relative regulations and decrees and will report to State Council on a regular basis. There are three types of monitoring and evaluation: (a) regular evaluation; (b) monitoring with prior notice; and (c) evaluating without prior notice.

(iv) Indicators

(1) A reduction in the number of water conflicts

(2) A reduction of water pollution and the improvement of water ecosystems, particularly wetland ecosystems

(3) Water productivity.

Integrated Strategic Plan for the Huaihe River Basin

Overall Social-economic Development Context of the Huaihe River Basin

The Huaihe River basin spans 35 prefectures and 181 counties in the five provinces of Hubei, Henan, Anhui, Shandong and Jiangsu. In 1997, the population was 160.43 million, with an average density of 594 persons/km2, which was 4.8 times the national average of 122 persons/km2. That made it one of the most densely populated regions in China.

The Huaihe River basin has a cultivated area of 183.26 million mu (12.22 million hectares). The main crops are wheat, rice, corn, yams, beans, cotton and rape. In 1997, the total agricultural output was RMB 284.4 billion (US$ 34.35 billion) or RMB 1,812 (US$ 218.8) per capita, which was significantly higher than the national per capita average of RMB 1,345 (US$ 162.4). These figures clearly indicate the basin's key role in national agricultural production. The main agricultural yield of the basin is grain. In 1997, the grain yield was 84.96 million metric tons, which accounted for 17.3 per cent of the total crop yield in the basin.

The Huaihe River basin is one of China's developing industrial regions. In 1995, the gross value of industrial output in the basin was RMB 763.4 billion and a GDP of RMB 541.2 billion (US$ 92.2 billion and US$ 65.4 billion, respectively). GDP per capita was only RMB 3,398 (US$ 410), which is lower than the national average of RMB 4,810 (US$ 581). The main industrial sectors in the Huaihe River basin are coal, power, food, light industry and textiles.

Opportunities and Threats for Water Resource Development from the Perspective of the Water Resources Council

Following the serious flooding in 2003, the State Council gave high priority to harnessing the Huaihe River. Investment has been greatly increased in order to guarantee the completion of important engineering construction works. The ability to prevent flood disasters is being promoted. Investment in environmental protection has also been increased, and the authority of the State Environment Agency has been strengthened with regard to controlling pollution in the Huaihe River basin. Integrated water resource management is taken into consideration during policy-making. The main challenges are flooding and water logging, water pollution and water scarcity.

A combination of climatic conditions and topography makes Huaihe River basin vulnerable to widespread, prolonged and devastating floods. While flooding event probabilities remain outside the control of society, protection against damage can be enhanced. Unfortunately, this is no easy task in the Huaihe River basin due, in part, to the climate and topography; however, an additional factor is that of the historical pattern of settlement in the plains and the extreme population density, which requires extensive infrastructure.

Surface and groundwater quality has been seriously degraded due to the lack of effective pollution control, combined with rising population and industrial operations, especially during the 1980s and 1990s. Causes of pollution from point sources include rapid urbanization, industrial development, rising population and the growth in the number of township and village enterprises and livestock operations in rural areas. Diffuse releases of nutrients and pesticides from agriculture are also becoming important sources of shallow groundwater and surface water contamination. Pollutants of concern recorded by SEPA from point sources include parameters for organics such as COD and BOD, phenols, ammonia nitrogen and inorganics such as heavy metals, while non-point source pollutant parameters include non-ionic ammonia, TP, TN and COD.

This dramatic decline in surface and groundwater quality continues despite efforts by regulatory authorities, including SEPA, and the Ministry of Water Resources, to improved control of the obvious sources of pollution such as state-owned enterprises as well as attempts to close small highly-polluting town and village enterprises. The current surface water quality status is such that most rivers and lakes fail to meet the standards required for the designated beneficial uses of the water. In 2002, the river stretches with water quality equal to, or worse than class IV was 69 per cent of the total river length.

In general, the water shortage problem in Huaihe basin is not particularly serious. However, despite the region's monsoon climate, with its greater variability in precipitation, water shortage problems occur in some dry years. Therefore, as a result of socio-economic development combined with water pollution, water shortage problems will become serious in the future.

Bibliography

Allchin, F.R.: *The Agriculture of Early Historic South Asia*, Cambridge University Press, U. K., 1995.

Bansil P C : *Agricultural Statistics in India*, CBS, Delhi, 2002.

Benghin, John C. : *Global Agriculture Trade and Developing Countries*, Manas, Delhi, 2005.

Bosworth, B.: *Capital Formation and Economic Policy*. Brookings Papers on Economic Activity, 1982

David Sadava: *Plants, Genes and Crop Biotechnology*, Sudbury MA, Jones and Barlett Publishers, 2003.

Dharamvir Hota: *Modern Agriculture and Plant Breeding*, Gene Tech Books, Delhi, 2007.

Dodds, John H.: *Plant Genetic Engineering*, New York, Cambridge University Press, 1985.

Edwards, C.A. *Sustainable Agricultural Systems*. 1990. Ankeny, Iowa: Soil and Water Conservation Society.

Jacquat Christiane: *Plants from the Markets of Thailand*, Bangkok, Duang Kamol, 1990.

Jasra, O.P. : *Environmental Agriculture*, Sarup & Sons, Delhi, 2002.

Kataria, T N : *Plant and Crop Physiology*, Pearl Books, Delhi, 2008.

Kumar, Arvind : *Environmental Pollution and Agriculture*, APH, Delhi, 2002.

Lakshmi, Narasaiah, M. : *Financing of Agriculture by Regional Rural Banks*, Sonali Pub, Delhi, 2008.

Lal, R., 1998. *Soil Quality and Agricultural Sustainability*. Chelsea, Mich.: Ann Arbor Press.

Mal, Pratap : *Infrastructural Development for Agriculture and Rural Development*, Mohit, Delhi, 2001.

Mellor, J. W.: *The New Economics of Growth*, Ithaca, Cornell University Press, 1976.

Mohanty, A.K. : *Entrepreneurship in Agriculture : Scopes and Opportunities*, Agrotech Pub, Delhi, 2011.

Murty, S. : *Faces and Phases of Agriculture and Industry in India*, RBSA Pub, Delhi, 2004.

Murugesan, R. : *Energy Use Efficiency in Dryland Agriculture*, Kalpaz, Delhi, 2010.

Nambiar P.T.N. : *Agricultural Statistics Techniques and Procedures*, Agrobios, Delhi, 2002.

Pareek O. P.: *Advances in Horticulture: Fruit Crops*, New Delhi, Malhotra Publishing House, 1993.

Parry M.L.: *Climatic Change, Agriculture and Settlements*, Dawson Folkestone UK, 1978.

Paul, Sharma, Vijay : *Glimpses of Indian Agriculture : Macro and Micro Aspects*, Academic Foundation, Delhi, 2008.

Purohit, S.S. : *Organic Farming : Biocontrol and Biopesticide Technology*, Agrobios, Delhi, 2008.

Rangaswamy R. : *A Textbook of Agricultural Statistics*, New Age Pub, Delhi, 2010.

Samvel, A P V : *Agri-Business Management*, Satish Serial Pub, Delhi, 2008.

Sastry E V Divakara : *Essentials of Agricultural Statistics*, Pointer, Delhi, 2007.

Sharma, J.P. : *Entrepreneurship in Livestock and Agriculture*, CBS Pub, Delhi, 2010.

Sharma, R.C. : *Diseases of Horticultural Crops: Fruits*, Indus, Delhi, 1999.

Sudharmai, C.R. : *Analytical Procedures in Soil Science and Agricultural Chemistry*, Agrotech, Delhi, 2004.

Swaminathan, MS : *From Green to Evergreen Revolution : Indian Agriculture: Performance and Challenges*, Academic Foundation, Delhi, 2010.

Tewari D.N. : *Tropical Forest Produce*, International Book Distributors, Dehra Dun, 1994.

Venkateshwara Rao, P. : *Dairy Farm Business Management*, Biotech Books, Delhi, 2008.

Vijaya, Kumar, T. : *Financing Agriculture by Primary Agricultural Co-operatives Societies in India*, Madhav Books, Delhi, 2010.

Index

❑❑❑